MILL AND CHLORINATION PLANT, CREIGHTON GOLD MINE, CHEROKEE COUNTY, GEORGIA

GEOLOGICAL SURVEY OF GEORGIA

S. W. McCALLIE, State Geologist

BULLETIN No. 19

SECOND REPORT

ON THE

GOLD DEPOSITS

OF

GEORGIA

BY

S. P. JONES

Assistant State Geologist

ATLANTA, GA.
CHAS. P. BYRD, State Printer
1909.

THE ADVISORY BOARD

OF THE

Geological Survey of Georgia

in the Year 1909

(Ex-Officio)

His Excellency, HOKE SMITH, Governor of Georgia
PRESIDENT OF THE BOARD

Hon. PHILIP COOK_____Secretary of State
Hon. J. P. BROWN_____State Treasurer
Hon. W. A. WRIGHT_____Comptroller-General
Hon. JOHN C. HART_____Attorney-General
Hon. T. G. HUDSON_____Commissioner of Agriculture
Hon. J. M. POUND_____Commissioner of Public Schools

LETTER OF TRANSMITTAL

GEOLOGICAL SURVEY OF GEORGIA,

ATLANTA, June 15, 1909.

To His Excellency, HOKE SMITH, *Governor and President of the Advisory Board of the Geological Survey of Georgia.*

SIR: I have the honor to transmit herewith for publication the report of Mr. S. P. Jones, Assistant State Geologist, on the Gold Deposits of Georgia. This report is the second report published by the State Geological Survey on this subject. The first report, which was issued in 1896, covered only a part of the gold deposits of the State, whereas this report covers all of the deposits; and, furthermore, gives all of the data now available on this very important branch of the State's mining industry. It is hoped that this report will stimulate gold mining in Georgia, which for the last few years has not received the attention which its importance warrants.

Very respectfully yours,

S. W. McCALLIE,

State Geologist.

IN MEMORIAM
WILLIAM SMITH YEATES
1856-1908

Having been intimately associated with Prof. W. S. Yeates for a number of years as Assistant State Geologist, and having subsequently succeeded him in office, I deem it a duty as well as a privilege to say here a few words concerning the character and work of this esteemed scientist. I would further add that I regard this an opportune place for these biographical notes, owing to the fact that one of his chief publications while in charge of the Geological Survey of Georgia was a report on the gold deposits of the State, and, furthermore, had his life been spared it was his intention to have been the senior author of this, the second report on the gold deposits of Georgia.

Prof. W. S. Yeates was born at Murfreesboro, North Carolina, December 15, 1856, where his father, Hon. Jesse J. Yeates, a United States congressman, then resided. He was graduated at Emory and Henry College, Virginia, in 1878, and in the spring of 1879 accepted a position with the United States Fish Commission. In 1880-81 he was employed in the Fisheries Division, 10th United States Census, which position he held until the winter of 1881 when he was appointed assistant curator of the United States National Museum, in charge of the collection of minerals and gems. The latter position he held until May, 1893, when he resigned to accept the position of State Geologist of Georgia. During Prof. Yeates' connection with the United States National Museum, he was Professor of mineralogy and geology in the Corcoran Scientific School of the Columbian University, Washington, D. C.

While State Geologist of Georgia, Prof. Yeates, in addition to his regular duties as director of the survey, did valuable work for the State at Buffalo, St. Louis, Jamestown, and other great national expositions, in displaying the mineral resources of Georgia. The selection and the installation of his exhibits could not be improved on, and they' invariably received the highest praise from those most competent to judge. This same remarkable talent for high-grade museum work is well illustrated in the Georgia State Museum, a worthy monument of his handiwork. Coupled with this genius for museum installation was the rarer faculty of determining mineral specimens at sight. His long experience in this line at the United States National Museum enabled him, not only to identify the species to which a mineral belonged, but in many instances he was also able to designate the locality from which the specimen was collected.

Personally, Prof. Yeates was a man of pleasing address and attractive manners, his cardinal virtue being honesty in its broadest sense. These qualities, together with his pleasant disposition and warm-heartedness, won for him a host of friends and admirers in his chosen profession. He was always conservative in his views and rarely ever undertook any work of importance that he did not carry to a successful termination. He was a man of broad and liberal ideas, but, at the same time, he was not by any means prone to accept scientific conclusions unless they were supported by reliable and trustworthy data.

From a professional standpoint, Prof. Yeates was especially noted for his advance views on practical geology. He was among the first geologists in this country to establish a state geological survey along purely economic lines. This same idea was strongly brought out in his arrangement of the State Museum and in his judicious selection of economic products for display at expositions. While the economic questions of geology always first appealed to him, yet, at the same time, the scientific side of the science was by no means ignored.

Prof. Yeates was a member of the Geological Society of America, The American Institute of Mining Engineers, and a number of other scientific organizations.

S. W. McCALLIE.

TABLE OF CONTENTS

LIST OF ILLUSTRATIONS

Gold Deposits of Georgia

CHAPTER I

INTRODUCTORY

General Considerations

This report is offered as a contribution to the economic geology of Georgia with the aim of furnishing to interested parties comprehensive and, as far as possible, definite ideas of the State's gold deposits and their mining possibilities.

Owing to the number of mines at present inoperative, some misapprehension exists regarding this branch of Georgia's mineral industry. It may therefore be well to outline certain past and present conditions.

Prior to the discovery of gold in California, the Southern Appalachian gold fields, of which the Georgia deposits form an important portion, received almost exclusively the attention of the gold miners of the United States. With the rush to California, and the subsequent discovery of gold in other western states, the Georgia mines were in large part abandoned by professional miners and lay idle for years, or were worked to a limited extent by local owners. The war between the States and the subsequent demoralized condition of business in the southern states for a number of years afterwards retarded further development for a considerable period. Later, with the revival of business in the South, companies were formed and mines on all the more important belts were

operated. In a number of cases the ore bodies were worked
below water level, and the refractory ores treated by the
cyanide or chlorination method. At some plants the ore was
milled, the free gold secured by amalgamation, and the sulphides concentrated for smelting. At many localities the
veins, especially where occurring as more or less parallel
stringers forming an ore-bearing zone, were sluiced with
hydraulic giants through large open cuts into mills situated
at lower levels. The depth to which the deposits could be
worked in this manner varied with the amount of the decomposition of the country rock. With some modifications, these
methods of mining, considering veins only, are in practice at
the present time.

Unfortunately in many cases the parties controlling and
directing the operations were eastern or southern business
men who had had little or no experience in gold mining and
errors of judgment were frequent in equipping plants. locat-
ing shafts, etc. Particularly regrettable was the erection, in
many instances, of mining plants, when neither the quality
nor the probable quantity of ore had received other than
the most meager tests. Another case of poor judgment
was in the use at a number of mines of Huntington mills
where the variety of ore mined could have been more satisfactorily and economically treated with stamp mills of the ordinary type. Failure under these conditions, after a longer or
shorter period of operation, followed and work ceased.

Another cause for many abandoned shafts and small mines
along all the gold belts is to be found in over-enthusiasm on
the part of small land owners on whose properties prospects
were located. Being farmers in most cases of limited means,
they were tempted to work, or to have worked, small stringers
or lean ore bodies that had much better been left undeveloped.

Also, in various localities veins were worked by the early miners to a depth where the sulphides were unoxidized and the ore ceased to be free milling. Unacquainted usually with any method other than amalgamation for extracting the gold, they were forced to cease operations and renew work at another locality. Numbers of this class of mines have been reworked under modern methods, but many are to be found as old caved-in and almost obliterated works.

It is not intended by the foregoing to convey the impression that some reason other than the absence of paying values accounts for all the idle mines in the several gold belts of Georgia. Taking an average of localities the world over where gold deposits have been worked, especially where underground work has been prosecuted, failures will bear a certain ratio, whatever that may be, to successful operations.

The above facts in regard to mines not in operation have been stated at some length, in order that undue weight may not be attached to a present condition of the industry that will appear fully in that portion of the report describing individual properties. It is also desired to impress upon property owners that the exploiting of trifling, and evidently unprofitable deposits, will result in injury to the future prospects of gold mining in the State.

Mining engineers familiar with the Georgia fields concur in the opinion that if deposits of a similar character existed in any of the more important gold producing states of the West they would have been much more extensively exploited. This is doubtless true; and a detailed survey of the territory confirms the belief that the Georgia deposits, especially in recent years, have not received the attention and tests they merit as judged by operations in many other gold yielding regions.

The conditions for mining the Georgia deposits are very favorable. Water and, in many cases, water power are abundant. Timber for construction purposes is plentiful, and can also be obtained cheaply for fuel if desired. The price paid for labor in the mining districts is reasonable. Transportation facilities are good; railways in a number of localities passing through or closely paralleling the gold belts, and being in no case farther than twenty-five miles distant. There are no extremes of climate to contend with, and the proximity of manufacturing cities like Atlanta, Chattanooga, Tenn., and Charlotte, N. C., renders the obtaining of supplies easy.

Loose and running ground is very rarely encountered. The veins occur in crystalline schists and gneisses, and while these are often found decayed to a depth of fifty feet or more, the residual material, though in some places mantled with a comparatively thin covering of foreign matter, is in place and firm enough to permit of easy working. Very unusual exceptions to these conditions may be encountered in certain cases where a pegmatite dike crosses a gold bearing vein at a small angle or runs for some distance by the side of it. The large amount of feldspar that these dikes frequently contain may become highly kaolinized to considerable depths, and under these conditions cause some inconvenience about water level.

History and Statistics

It is thought by some that gold was mined to a limited extent in Georgia by Spanish explorers when De Soto's expedition passed through the State. Several traditions also credit the aboriginal tribes with having discovered and secured placer gold especially in the Dahlonega district of Lumpkin county. It is probable that the Indians found nuggets in this region, and prizing them, doubtless learned to search for them at favorable localities. But the suggestion by some writers that they washed gold out of the stream beds seems unlikely when the difficulty of getting down to the bed rock of the average Appalachian stream carrying its large load of sand and rock boulders is considered.

In 1829 gold was discovered nearly simultaneously in the Nacoochee Valley region of White county and near Dahlonega in Lumpkin county.[1] Many of the placer deposits of these two regions, occurring along small branches and creeks, were exceedingly rich and easily worked, and active mining followed the discovery. From these two centers mining operations gradually spread to other localities.

By 1838 the production of gold in Georgia and North Carolina had become of sufficient magnitude to warrant the establishment by the United States government of a branch mint in that section. This mint was located at Dahlonega in Lumpkin county, and operated from the date just mentioned until the year 1861, the time of the secession of Georgia from the United States. The report of the United States Treasury Department shows a total coinage at Dahlonega during that period of $6,115,569.00. From statistics from the Director of the mint and other sources it is estimated that the entire production of gold in Georgia from early discoveries to 1909 has amounted to about $17,500,000.00. The yield has fluctuated greatly in different years, the largest output per annum having occurred prior to the Civil war before the more important placers were exhausted.

Before the war between the States, while the major portion of the production was derived from placer deposits, vein mining was not neglected. The operations, however, were limited to the mining of free milling ore, and when the level of ground water was reached the mines were often abandoned.

---

1. It is claimed that gold was discovered prior to this date in McDuffie county in the vicinity of the Columbia mine.

Fluker, W. H., Trans. Am. Inst. Mining Engineers 1903, Vol. XXXIII, pp. 119-125.

Stamp mills of small capacity, not infrequently manufactured by the miners themselves out of oak timbers with iron shod shoes and dies, were used for crushing the ore and in some cases arrastres were employed.

During the period of the war above mentioned and for several subsequent years, mining was practically at a standstill. With the resumption of business in the South, gold mining was renewed and with the gradual introduction of modern methods deep mining and the treatment of refractory ores became a permanent industry.

CHAPTER II

A CONSIDERATION OF THE DIFFERENT TYPES OF DEPOSITS

INTRODUCTION

The three principal types of gold deposits recognized in Georgia are: (1) vein deposits; (2) placer deposits—consisting of beds of auriferous stream gravel, both ancient and modern, which include as a sub-class, gulch and hillside deposits consisting of soil and decomposed rock brought from higher levels by rain-wash and the action of gravity and having usually a more or less irregularily occurring sub-stratum of angular or slightly sub-angular gravel;[1] (3) auriferous saprolites or decomposed rock in place. Partially disintegrated quartz veins and numerous tiny stringers of quartz are usually present, and in many of these saprolite deposits there are found distinct zones of more or less parallel veins or stringers of gold bearing quartz intercalated with the rotten rock which at depth in undecomposed material would be classed and mined as a vein deposits.

VEIN DEPOSITS

Several varieties of veins are to be noted. The predominant types are fissure veins (the term *fissure vein* is not here restricted, as is frequently the usage with gold miners, to

---

1. A similar class of deposits has been termed "Colluvial deposits" by some writers.

veins cutting the trend of the inclosing rocks at an angle),
conforming in the main to the trend of the inclosing schists
and gneisses which is usually northeast and southwest.   Thin
stringers running out from the main vein for short distances
into the wall rock are common and not an unusual feature in
sinking on a solid quartz vein is to find it breaking up into a
number of thin parallel bands with intercalated wall rock
making up in some instances 50 per cent. of the vein.   In such
cases, however, the outlines of the vein as a whole are usually
preserved and the intercalated rock is frequently mineralized
and generally carries a percentage of the values.   Extreme
cases of this character are sometimes exposed to observation
in prospect pits where a seam, in some instances several feet
wide, will be noticed in the decomposed rock, differing slightly
in character from the balance and having fairly definitely
defined walls, but showing apparently no vein quartz.   Close
observation, however, will usually detect some thin laminæ
of that material.

The veins conforming to the schistosity of the enclosing
rocks pinch and swell both horizontally and vertically causing
one of the greatest difficulties attendant on mining this class of
deposits.   Bodies of ore fifteen or twenty feet in thickness
may occur at some points while a short distance away on the
same level a small stringer or seam of quartz may be all that
represents the vein.   This necessitates not infrequently the
driving of a drift at much expense through the country rock
along a quartz stringer until a workable portion of the vein
is again encountered.   These conditions also render it difficult for the mine superintendent to calculate the amount of
ore available and to utilize at all times the full capacity of his
plant.   At some mines, however, it has been noted that the
pinches and swells occur with a fair degree of regularity, as
at the Franklin, or Creighton, mine in Cherokee county, where

the ore bodies are said to average sixty or seventy feet in linear extent. It has been emphasized by one of the prominent mine managers that it is advisable in working such deposits to have the levels as far apart as practical in order to minimize the expense of driving through tough schists and gneisses.

In connection with the veins, conforming as a class with the schistosity of the country rocks, a type of deposits is to be described that may be designated as gold bearing zones. These consist, as the name implies, of auriferous zones having in most cases the trend of the formations in which they occur and containing many more or less conformable quartz stringers or lenses of quartz with country rock. They vary from a foot or two to several yards in thickness. Frequently, several more or less parallel zones, ranging from a few to twenty or thirty feet in thickness, occur closely associated, the distribution as a whole forming a major zone in some cases several hundred feet in width the strike of which may, in certain instances, be followed for miles. In these large zones the paying values, as would be expected, are limited to bands and restricted areas. A good example of one of these gold bearing zones is to be seen at the Jones mine, or Lot 10, in Nacoochee Valley, White county. It is along the outcrop of these gold bearing zones that the saprolite deposits previously mentioned frequently occur. And, as massive bodies of auriferous quartz are sometimes found along with the stringers, the zone deposits grade on the one hand into saprolite deposits and on the other into ordinary vein deposits, the upper decomposed portion being worked by hydraulicking as a saprolite deposit while the bunches of quartz are mined from shafts sunk in the cuts formed by the sluicing operations. The gold bearing zones are frequently designated locally as gold belts. As the major portion of the Georgia deposits, considered as a whole, occur in definite geographical belts, it would seem preferable to restrict that term to these larger divisions.

In addition to interfoliated veins, others cutting the schistosity of the enclosing rocks at various angles are not uncommon. It is generally believed that these are more uniform and persistent than the other class. Swells and pinches and other irregularities of structure, however, are frequently met with. These non-interfoliated veins are quite common in the McDuffie belt and a number of them have been good producers.[1]

Some of the auriferous veins of the Georgia deposits may be replacement veins. Owing to the limited number of accessible underground works that had penetrated deep enough to permit of observation in rocks not partially decayed, studies in regard to this proved rather unsatisfactory. At the Seminole Gold and Copper mine in Lincoln county, it is thought that the ore bodies owe their origin, in part at least, to replacement of the country rock along lines of intense shearing. A discussion of this particular occurrence will be found in the chapter on the description of individual properties.

In size, the veins in Georgia, as in other sections of the Southern Appalachian gold fields, vary from a few inches to over twenty feet in thickness. The outcrop of many of the large veins is very noticeable and shows them in some cases holding their width quite uniformly for a number of rods. Unfortunately, the most of the largest veins that have been tested show too small value to admit of working. They have not been sufficiently exploited, however, to establish this as a rule.

---

1. Fluker, W. H., Trans. Am. Inst. Mining Eng., Vol. XXXIII, 1903, pp. 119-125.

It may be of interest to note in this connection that numbers of very small veins only a few inches in thickness have been mined at various localities that contained astonishingly large amounts of gold. One would feel a hesitancy in stating the yield from some of these little stringers, but reliable data are obtainable in many cases. Veins in Georgia large enough, however, to warrant the erection of a mining plant with the prospect of long continued operations should not be expected to yield what would be ordinarily considered very high values. An average in such cases of from $6.00 to $15.00 per ton may be considered good.

The principal gangue mineral of the veins is quartz. The most frequently occurring metallic sulphide is iron pyrite. Chalcopyrite and galena come next. The former is probably more widespread in occurrence than the latter. Sphalerite has been noted at only a few mines. It is fairly abundant at the Seminole mine in Lincoln county and occurs in small amounts in a vein on the Currahee property in Hall county. Arsenopyrite has been observed at only one or two localities. For an extensive list of the various minerals found in the gold bearing veins of the Southern Appalachian region, reference may be made to Dr. Geo. F. Becker's valuable paper on the Gold Fields of the Southern Appalachians.[1]

Both the gold and other constituents of the veins are found at many localities impregnating the wall rocks, and in some cases enough of the walling carries sufficient values to materially increase the supply of millable ore. In some instances deposits have been exploited that should probably be classed as zones of impregnated country rock rather than as true veins.

---

1. Becker, G. F., Gold Fields of the Southern Appalachians: Sixteenth Ann. Rept. U. S. Geol. Surv., Part 3, 1895, pp. 272-281.

In connection with the subject of vein minerals, it is of interest to note that, while the presence of one or more metallic sulphides, or mineralization of the veins, to use the miner's term, is generally highly favorable to good values, yet a number of small veins or stringers containing apparently little or no metallic sulphides have been found to be quite rich in gold. It is also curious to note that, at certain localities, within a mile or so of gold mines, distinct veins occur carrying so high a percentage of iron pyrite that they are worked as pyrite deposits, yet they contain very little, if any, gold.

In regard to the occurrence of the gold in the veins it may be stated that, as in most auriferous quartz veins, the gold is largely associated with iron pyrite. The extent of the association, however, seems to vary in different localities and in individual veins. In some cases a considerable percentage of the gold, below the zone of decomposition of the sulphides, is found free in the quartz and can be secured by amalgamation. In other cases, an assay of the milky white vein quartz from which the sulphuretted portions have been culled will show only a very small percentage of the total value.

The distribution of the gold in the veins is rarely uniform, but varies both laterally and vertically, and the values are frequently found in ore chutes, generally pitching along the strike of the veins at a rather steep angle.

There seems to be no ground for the frequently expressed belief that the veins will grow richer in depth. While in some exploited deposits the values have increased with depth in others they have decreased. Indeed, in the upper altered portions of the veins, a considerable part of the metallic sulphides has been carried away in solution leaving the vein quartz more or less porous. Tests based on a given weight of this ore, as is the usual practice, should show higher values than on similar portions of the vein lower down where less bulk of the ore would weigh the same. This assumes that very little gold from the upper decomposed portion of the vein has been carried in solution and redeposited lower down. Owing to the well established facts in regard to the secondary enrichment of copper veins, and it being recognized that gold is slightly soluble in circulating waters holding certain compounds in solution, many mine operators have hoped to find much better values below the level of ground water than above. It is probable, however, in the Appalachian region that removal of the upper portions of the veins by erosion has been too rapid to permit of much solution of the gold by circulating waters with subsequent deposition partly in the oxidized zone and also in the region of the unaltered sulphides. As the original outcrops of many of the more continuous veins were doubtless several thousand feet above the present level, if much solution with redeposition had taken place the results of concentration would have been cumulative, as the level of the zone of decomposition was gradually lowered. Movements in the earth's crust causing a more rapid rate of erosion than the average might have partially or entirely removed such accumulations, but the fact that at exploited deposits no prevailing marked increase in values near water level has been noted, argues against secondary enrichment.

In concluding a description of the character of the vein deposits it should be added, that though at a few localities silver and copper have been obtained in commercial quantities with the gold, profits may generally be expected only from the yield of the gold.

PLACER DEPOSITS

The placer deposits bear a close relation to the vein deposits, that is, in sections where the veins are most numerous and carry good values the placers have been found to be proportionately important. A few apparent exceptions should be made to this general statement. In the McDuffie belt no placer deposits of any importance have been located. This belt is only about twenty miles north of the upper edge of the Coastal Plain in a portion of the Piedmont Plateau that approaches much nearer to a condition of base level than any other of the gold bearing sections. The physiographic conditions of steep slopes with narrow gulches and V-shaped stream gorges favoring the rapid concentration of gold in placers are absent. Also, the conditions for prospecting for whatever accumulations of gold that may be along the stream courses are more difficult than in sections where steeper grade has caused the small creeks and branches to cut through and expose gravel beds along their banks. In Gilmer county, the White Path mine is a rather noted placer deposit seemingly isolated from any well defined gold belt. A large amount of gold has been obtained for a distance of about a mile along a small stream and in bordering gulches. According to reports the largest nuggets that have been found in the State were secured years ago at this locality, two having been mined that weighed about eight hundred pennyweights apiece. No returns have ever been reported from veins. Whether the gold of the placer was derived from numerous small stringers in the country rocks that have escaped observation, or whether its source was from veins that pinched out at the present level of the land, is a matter of conjecture.

As stated at the beginning of this chapter the auriferous gravel deposits may be classed as both ancient and modern. It is difficult, however, to draw a line between the two classes at many localities.

The valleys of the larger streams, where they run with or across the gold belts, are frequently underlain with quite heavy beds of thoroughly water-worn gravel overtopped with sand and clay. In some cases tenaceous blue clay is mixed with the gravels and forms a layer several feet in thickness immediately above the gravel beds. In the lowest parts of the valleys the gravels are usually near the present level of the streams and are to be found thinning out at varying heights along the slopes of the valleys. The different auriferous gravel deposits so far located are associated with present drainage systems and while placer gold can be obtained at some localities as high as thirty or forty feet above the streams, yet the bulk of the gold that has been mined has come from lower levels. A good type of these deposits is to be seen in the broad valleys of the streams in Nacoochee Valley in White county. These broad deposits owe their origin to a flooded condition of the streams during a period of subsidence. The changes of level are probably to be referred to the Lafayette or the Columbian periods.

Many of the richest placer deposits have been found along the small streams feeding into the larger valleys and in some cases very rich finds have been made in gulches or dry hollows, as they are termed, near the smaller streams. These latter deposits grade into the modern deposits previously mentioned as formed by rain-wash and gravitative action. The beds of some of the streams of the Dahlonega belt, notably that of the Chestatee River, have been found to yield sufficient gold to warrant dredging operations. These stream-bed deposits should, of course, be classed as recent deposits, the gold no doubt having been derived largely from the old gravel deposits through which these streams and their smaller tributaries have cut their way.

Regarding the future of the true placer deposits, it should be stated that the great majority of them have been located and a large percentage extensively worked under old methods, many having been re-worked a number of times. Some of the old placers, where they can be re-worked cheaply and thoroughly, may yet yield moderate profits. In some localities also, where the streams run through large flood plains and the overburden of the gravel is heavy with the grade of the streams very slight, considerable areas are to be found that have never been worked, or the worked portion is limited to small strips along the stream margins. There is the possibility of such areas yielding considerable quantities of gold. Dredging would seem to be the most feasible method of working these broad flood plain deposits.

The deposits of gold in the beds of streams, in and for some distance below auriferous belts, have been worked with dredge boats at a number of localities. Where these boats were operated by parties experienced in dredging work they generally yielded good profits. It is rather difficult to estimate what portion of this class of deposits has been worked out, but that there is a considerable field for continued operations seems fairly certain.

Black Sand

In recent years some investigations have been undertaken, especially in the Dahlonega district, to ascertain the probable value of the gold contents of the black sands of the streams flowing through auriferous areas. Dredging operations in the Chestatee River near Dahlonega have demonstrated the presence of considerable amounts of these sands in the bed of that stream. At the time of visit a dredge was in process of equipment for securing black sand. As this had not commenced operations, samples of the sand for study could not be obtained from the bed of the stream. Several samples, however, were secured from sand bars on this stream and from other streams and from placer mines in the Dahlonega region. A sample of black sand was also taken from the Loud mine in White county. The samples were secured by panning and tested without further concentration. A small percentage of ordinary siliceous sand was therefore included in each sample. Since field work was completed, Prof. L. M. Richard, of the North Georgia Agricultural College, at Dahlonega, kindly collected and furnished a sample of black sand from the bed of the Chestatee River. This sample showed a slightly higher degree of concentration than the others.

In most of the samples examined free gold in the form of small water-worn particles was present in varying amounts. The two most abundant constituents of the sand from all the localities were seen to be magnetite and ilmenite. By far the larger proportion of the dark colored grains was found when tested to be one or the other of these two minerals. In some samples magnetite was in excess and in others ilmenite. Small crystals and fragments of garnets were found to be present, but in considerably smaller amounts than either of the minerals previously mentioned. In some of the black sands small particles of either cyanite or sillimanite are present in noticeable amounts.

In testing the sands collected by the Survey for their auriferous contents a sample from each locality was pulverized sufficiently fine to admit of passage through an eighty mesh sieve and ground thoroughly for several hours with mercury. The mercury and sand were then separated and each assayed. In the case of the sample furnished by Prof. L. M. Richard, a preliminary assay was made from a portion as if testing an ordinary gold ore. A portion was also pulverized and amalgamated in the same manner as in the case of the other samples and the gold extracted by the mercury assayed. In addition, an assay was made of a part of the material separated from the balance by means of a magnet.

The results of these tests are given below:

Black Sands from Sand Bar in the Chestatee River near the Briar Patch Mine

Assay of gold extracted by mercury_$9.00 per ton

Assay of sand after amalgamation__ 0.00 per ton

Black Sand, Ritchie Placer Mine, near the Chestatee River

Assay of gold extracted by mercury $35.96 per ton

Assay of sand after amalgamation_ 0.00 per ton

Black Sand from Bed of Chestatee River at Briar Patch Mine

(Collected by Prof. L. M. Richard).

Preliminary assay _____$112.27 per ton

Assay of gold extracted by mercury 67.18 per ton

Assay of material separated by

magnet _____ 0.00 per ton

Black Sand, Loud Mine, White County

Assay of gold extracted by mercury_____$3.31

Assay of sand after amalgamation_____ 0.00

The results obtained from the tests of all the samples except the one from the bed of the Chestatee River, would tend to show that the entire auriferous contents of the sands are in the form of free gold. That, in portion at least, in the samples assayed, it was in the form of water-worn grains was shown by preliminary panning tests. It has been suggested by some investigators of black sands in other regions that gold occurs as a coating on the grains of iron oxide.

SAPROLITE DEPOSITS

The saprolite deposits of the Southern Appalachian region form a rather unique class of American gold deposits. The rocks of this section are deeply decayed at many localities. The decayed product is found in place, there having been no invasion of the region by glaciers during the period when removal of decomposed rock material by ice erosion was taking place in more northern localities.

As previously stated, in these auriferous bodies of decomposed rock more or less disintegrated quartz veins and many small stringers of quartz are usually present. The gold, generally freed from sulphides, is found both in the quartz and in the rotten rock. In some cases the veins of quartz may be very small, occurring as numerous tiny stringers and seams, and in other cases some individual veins may be of sufficient size to form workable ore bodies. In the latter case, as before mentioned, these larger veins are sometimes mined independently as ordinary vein deposits, as the surrounding saprolite material grades into undecomposed rock below and can no longer be worked by hydraulicking. Plate III, Fig. 2, gives a view of numerous quartz stringers in partially decomposed rock in the east cut of the Findley mine at Dahlonega. The saprolite deposits have been extensively worked by sluicing the material through flumes to mills located at some point below the level of the cuts thus formed. A portion of the gold is generally saved by amalgamation in the sluice boxes and the fine portion of the sluiced material is conducted to ore bins behind the batteries of the mills in which it is to be treated. The larger pieces of quartz and partially decomposed rock are removed by grizzlies or some similar contrivance before the current reaches the ore bins, and this material is frequently crushed at the more extensive plants in a separate mill.

Immense cuts have been made in the hills about Dahlonega in working these saprolite deposits. Plate IV, Fig. 1, shows a view of a portion of the cut of the Barlow mine which is over half of a mile in length. At this great cut there was formerly a mill at either end, the material being sluiced both ways.

Although these deposits are largely worked by washing, and have been described under the heading of placers by several writers, they are not to be confounded with true placer deposits in which the gold has been concentrated by the transporting power of water. In saprolite deposits a certain amount of concentration of gold has taken place in two ways: (1) by the removal of some portions of the rock ingredients in solution while the insoluble, or only slightly soluble gold, is left behind, a given amount by weight of the saprolite would contain a greater amount of gold than the same quantity of the original unaltered rock; (2) also, gold freed by surface weathering from the quartz stringers or from the rock on account of its high specific gravity would tend to settle into cracks and crevices and work its way downward for a short distance into the more or less porous mass of the upper part of the rotten rock.

While probably the greater portion of the gold that was in the denuded part of the saprolite has gone into alluvial placer deposits or been widely distributed by the larger streams, yet a certain amount of concentration in the upper part of the deposit has doubtless taken place as above pointed out. Where weathering has been long continued on gentle slopes the amount of gold left would be in excess of that at localities where the decaying surface was removed more rapidly.

Owing to the very limited number of localities where any work was in progress on the saprolite deposits during the preparation of this report, little opportunity was offered to ascertain what percentage of the gold has been saved and what lost under the method of mining employed. Different observers seem to have arrived at different conclusions in regard to this. Dr. Becker,[1] in a paper previously referred

---

1. Becker, G. F., Gold Fields of the Southern Appalachians: Sixteenth Ann. Rept. U. S. Geol. Survey, pt. 3, 1895, p. 301.

to, in discussing economic considerations, says: "The present method of working the saprolites is very wasteful and yields insignificant returns. If means can be found to save the rusty gold much money may be made." In a rather detailed description of the mines about Dahlonega in Lumpkin county, Ga., in Bulletin No. 10 of the North Carolina Geological Survey, by Messrs. Henry B. C. Nitze and H. A. J. Wilkens.[1] the following is noted: "Despite many inquiries amongst local millmen and others, we could hear no reports of losses in amalgamation resulting from so-called rusty gold. A loss of this nature was in a few cases ascribed to the finely divided or flaky condition of the gold."

In numerous panning tests of saprolites made at different localities the writer rarely encountered any gold that under a hand lens appeared other than clean and bright. At one or two localities in the Dahlonega region a small percentage of the gold thus obtained was seen to be partially or entirely covered with some black substance. Small particles of gold coated with metallic oxides or sulphides might, however, readily escape observation in panning. In panning oxidized ore several cases were noted where auriferous pyrite had passed over into compact iron oxide without the included particles of gold being set free. The values of material of this class in a saprolite or placer deposit would, of course, not be saved by amalgamation unless the ore was finely pulverized before treatment. It is also stated by some mine managers that, at certain deposits, gold appearing clean and bright does not readily amalgamate.

The fineness of the Georgia gold is high at nearly all localities where it has been mined. The Loud mine in White county is an exception to the general rule. The gold there

---

1. Gold Mining in North Carolina and Adjacent Appalachian Regions: Bull. No. 10, N. C. Geological Survey, 1897, p. 114.

has been reported as 0.800 and even lower. At the placers along Coosa Creek in Union county it is claimed that the gold has a fineness of 0.980. Dr. Becker, in the paper recently quoted from, places the average at something like 0.950.

CHAPTER III

GEOGRAPHICAL AND GEOLOGICAL DISTRIBUTION OF THE DEPOSITS

The gold deposits of Georgia are found in a portion of a broad zone of country stretching from near the centre of Alabama northeastward into Maryland, and lying principally southeast of the Blue Ridge mountains. In Georgia, this zone takes in the greater portion of the Piedmont Plateau and a part of the physiographic provinces of the Appalachian mountains. Auriferous areas in which are found gold bearing quartz veins and other forms of deposits occur throughout this region, generally as more or less parallel belts of relatively narrow lateral dimensions, but they are found at some localities as small isolated areas or patches.

In Georgia, the larger portion, by far, of the auriferous areas occur in narrow well defined belts, and show, as will be seen by reference to the map accompanying, as bands of varying width running nearly parallel in a northeast and southwest direction. In addition to the belts a few isolated areas are found in the same section in which the belts occur. It will be seen from the map that all the deposits are north of the Fall Line, a line formed by the junction of the Piedmont region and the Coastal Plain. In Georgia, this line passes approximately through the cities of Columbus, Macon and Augusta. The belts parallel in a general way the axis of the Blue Ridge mountains and the larger portion of them lie southeast of this axis. These individual belts are here described.

THE DAHLONEGA BELT.—The Dahlonega belt enters Georgia from Alabama and passes through Haralson and Paulding counties, the northwest corner of Cobb and the southeast corner of Bartow counties, traverses Cherokee county and the extreme northwest corner of Forsyth county and from thence passes through Dawson, Lumpkin, White, Habersham and Rabun counties into Macon county, North Carolina. The Dahlonega belt has a length in Georgia of about one hundred and fifty miles and varies in width at different localities from two to six miles.

THE HALL COUNTY BELT.—The Hall county belt commences in the northern part of Fulton county, runs northeast through Milton county, the extreme southeast corner of Forsyth, the northwestern part of Gwinnett, and from thence through Hall, Habersham and Rabun counties into North Carolina. The length of this belt in Georgia is about a hundred miles.

THE McDUFFIE COUNTY BELT.—The McDuffie county belt commences in the northeast corner of Warren county, runs northeast through the northern portion of McDuffie county, the extreme southeast corner of Wilkes county, and from thence through Lincoln county to the Savannah River. Its continuation is probably found in South Carolina in the North Carolina belt. The McDuffie belt is nearer the Coastal Plain than any other known belt, its distance from the Fall Line being about twenty miles. The length of this belt in Georgia is about thirty miles with an average width of about two miles.

THE CARROLL COUNTY BELT.—The Carroll county belt commences in the western part of Carroll county and, running northeast, passes through a corner of Douglas and thence traversing a portion of Paulding and Cobb counties joins the Dahlonega belt at the northern edge of the last named county.

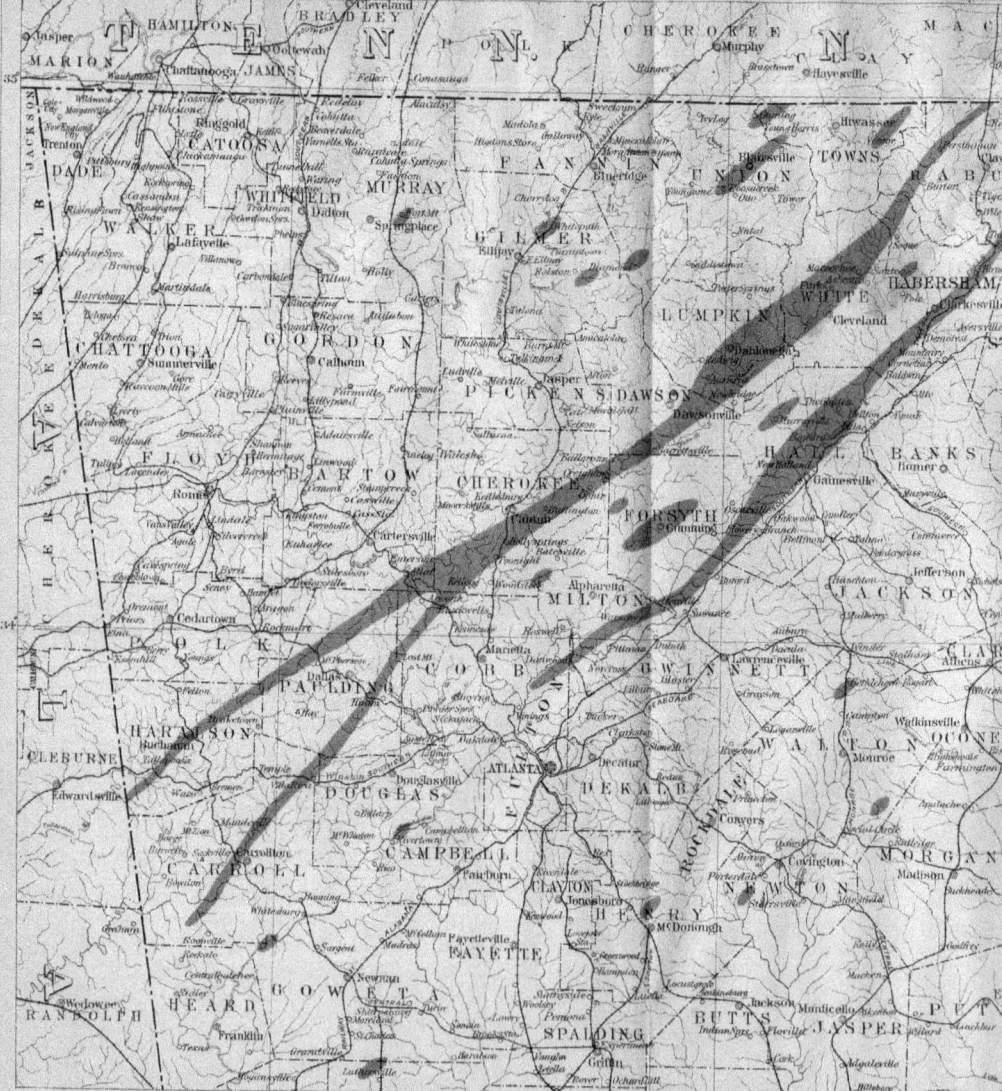

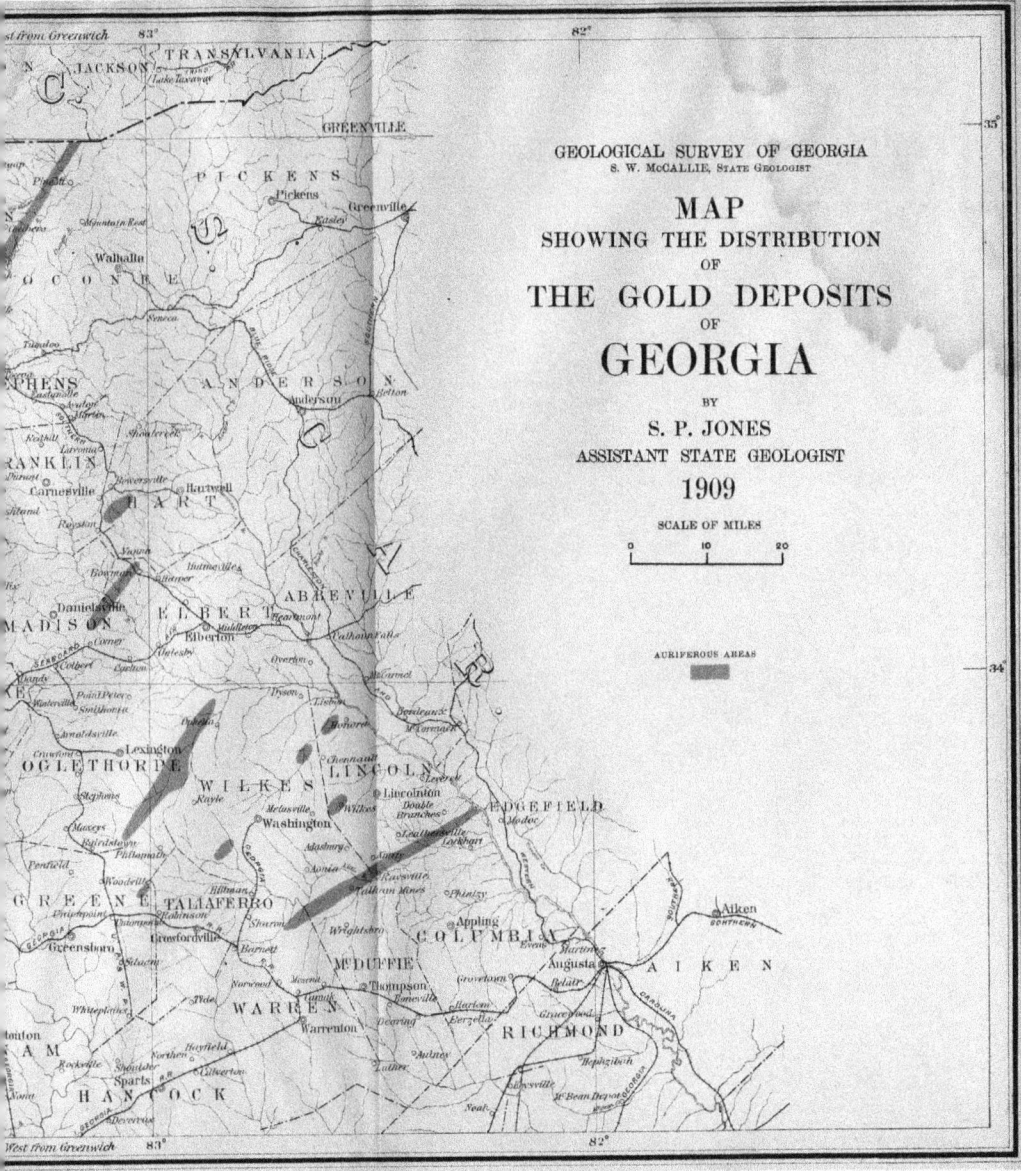

GEOLOGICAL SURVEY OF GEORGIA
S. W. McCALLIE, State Geologist

MAP
SHOWING THE DISTRIBUTION
OF
THE GOLD DEPOSITS
OF
GEORGIA
BY
S. P. JONES
ASSISTANT STATE GEOLOGIST
1909

SCALE OF MILES

0   10   20

AURIFEROUS AREAS

As thus outlined, its length is about fifty miles and its greatest width does not probably exceed two miles.

THE OGLETHORPE COUNTY BELT.—The Oglethorpe county belt runs northeast through the eastern part of Oglethorpe county. Its length is something like twenty-five miles and its width is about the same as that of the Carroll county belt.

THE MADISON COUNTY BELT.—The Madison county belt occurs in Madison and Elbert counties, and extends from a point a few miles northeast of Comer in Madison county to a point about three miles northeast of Bowman in Elbert county. Its total length, as far as it has been traced, is only about ten miles.

All of the belts, the locations of which have thus far been described, lie, excepting a small portion of the Dahlonega belt in Rabun county, south, or southeast, of the crest of the Blue Ridge mountains, the most eastern range of the Southern Appalachians. Several belts of minor dimensions are found north of the crest line of the Blue Ridge.

THE GUMLOG BELT.—The Gumlog belt runs from a point a little south of the Gumlog mine in the northern part of Union county northeast through the northwest corner of Towns county to the Warne mine immediately beyond the State line in North Carolina. Its length in Georgia, as far as it is known, is about eight miles.

THE COOSA CREEK BELT.—The Coosa Creek belt runs from near the headwaters of Coosa Creek in Union county northeast to a point in the neighborhood of the town of Young Harris in Towns county. Its length is about fifteen miles.

THE HIGHTOWER CREEK BELT.—The Hightower Creek belt runs from near Mountain Scene on the headwaters of Hiawassee River in Towns county northeast to within a few miles of the Georgia-North Carolina line. Its length is about ten miles.

ISOLATED AREAS.—In addition to the different well defined belts above mentioned, a number of localities are to be noted where gold has been found in isolated areas. The relations, however, of some of these deposits indicate that future prospecting will probably connect a portion of them at least into belts with the usual northeast-southwest trend. The counties in which are to be found the more important deposits of the class just described are: Fannin, Gilmer, Lincoln, Hall, Cherokee, Meriwether, Forsyth, Wilkes and Murray. A little gold has also been mined in Hart, Walton, Coweta, Campbell and Newton counties. It has also been found in very small amounts in Henry, Clarke, and one or two other counties.

Reference to any geological map of the State will show that all the gold deposits of Georgia, geologically considered, occur in the large area designated as "Igneous and Metamorphic." This area is also commonly spoken of as the Crystalline area, and is composed of rocks that have, for the most part, been profoundly altered by dynamo-metamorphism and converted in many cases from originally simpler types to gneisses and schists of varying complexity of structure. The geology of this region, together with the character of the rocks and the relation of the auriferous veins to them, is discussed in the chapter immediately following.

CHAPTER IV

GEOLOGY AND GENESIS OF THE DEPOSITS

GEOLOGY

The geological formations in which the gold bearing veins occur are the most ancient in the State, and among the oldest of all the formations of the North American continent.

In Georgia, the rocks of these formations have been tentatively classed as pre-Cambrian, though at the present time work is in progress in portions of the Crystalline area by the United States Geological Survey to differentiate and assign them to the various sub-divisions of time. A map by Mr. Arthur Keith of the Federal Survey accompanying this report shows the results of work of this character in the Dahlonega district. In the northwestern part of the region, among these ancient rocks highly altered original sedimentaries are to be found that are usually referred to the Cambrian period. Unaltered, or only slightly altered, intrusive diabases of a later age than Cambrian occur at numerous localities, but their total area is insignificant compared with that of the entire region.

A very casual study of the rocks of the Crystalline area will show that the majority of them have been subjected to great shearing stress. Occasionally bodies of granite, probably younger than the neighboring formations, can be found that appear to have been subjected to only a limited amount of pressure, but the major portion of the rocks are gneisses and schists often showing a highly contorted as well as sheared structure. While widespread regional metamorphism is characteristic of the area as a whole, yet at some points, lines are found along which shearing forces appear to have been especially concentrated. The ore bodies of the Seminole mine in Lincoln county are located on one of these lines of intense shearing.

The auriferous veins were probably formed during the closing period of the last earth crust movement that produced important changes of structure in the rocks. The evidence that their age does not exceed this is found in the fact that the majority of them do not show any large amount of faulting or crushing. All of the veins may not be of the same age and if more definite evidence in regard to age relations was obtainable it might be found that several series of veins are represented in the area. Most geologists who have written on the Southern Appalachian fields arrived at the conclusion that the greater part, at least, of gold deposition antedated Cambrian times. Becker, in discussing this question, says: "The greater part of the gold I believe to have been deposited at the close of the great volcanic era, or during the Algonkian. In the Carolina belt, this conclusion seems inevitable. and I know of no good ground for supposing the ore and the granitic dikes of the South Mountain area and of Northern Georgia to be younger. Gold deposition was seemingly renewed with diminished activity after the Ocoee and Monroe beds were laid down."[1]

Mr. L. C. Graton, in a paper on a portion of the gold deposits of the Carolinas, in concluding a short discussion as to the probable age of the deposits, says: "All available evi-

---

1. Becker, G. F., Gold Fields of the Southern Appalachians: Sixteenth Ann. Rept., U. S. Geol. Survey, pt. 3, 1895, p. 261.

dence points to pre-Cambrian as the probable age of the deposits.''[1]

A period of volcanic disturbance, extending over certain areas in the Appalachian region, may be referred to Algonkian times and was probably accompanied by fissuring and the formaton of quartz veins. Also, while the Paleozoic may have been a comparatively quiet period over the greater part of the Atlantic slope, yet there are certain considerations in regard to the auriferous quartz veins in northern Georgia that render it difficult to class the larger portion of them as pre-Cambrian. Reference to the map accompanying this report will show that, at the corners of Paulding, Bartow and Cherokee counties. the Dahlonega gold belt is within about three miles of the Paleozoic area of Northwest Georgia and for fifteen miles, or more, both northeast and southwest from this locality, it is very close to Paleozoic rocks. The Gumlog belt in Union county is also in close proximity to rocks in North Carolina that have been mapped by Keith as Cambrian. In both of these gold belts, at the localities mentioned, numerous quartz veins can be found consisting of solid, massive, milky quartz showing little evidence of shearing or crushing. Taking into consideration the shearing forces that must have been operative in producing the slates, schists and other metamorphosed rocks found in the eastern and southern edges of the Paleozoic area in Georgia, it is difficult to understand how pre-existing quartz veins a few miles distant could have escaped marked metamorphism. Some of the gold deposits of Georgia that occur as isolated areas are found in rocks that from their position can not reasonably be referred to older formations than the Cambrian. The principal vein of the Cohutta gold mine on Cohutta Mountain in Murray county occurs in a metamorphosed sediment, probably belonging to

---

1. Bull. U. S. Geol. Survey No. 293, p. 74.

the Ocoee. A fine grained, bluish schist or slate is to be seen a few yards from this auriferous quartz vein, but an open cut from which the vein has been mined is in a sheared arkosic sandstone or conglomerate. In North Carolina, ancient volcanics have been found in the region of the gold deposits and referred to Algonkian times, but in studying the rocks associated with the Georgia deposits, nothing indicating an original surface flow has been observed. It will be noticed that one of the rocks described in this chapter shows in the ground mass a fairly well defined coarsely spherulic structure, but the specimen is readily classed with acid porphyritic dike rocks. The only direct evidence in regard to age relations is the presence, at some localities, of certain diabase dikes (usually assigned to Jura-Triassic time), intersecting the veins as younger formations. One of these dikes is to be seen cutting the Franklin vein in shaft No. 3½ of the Creighton mine in Cherokee county.

Genesis of the Deposits

Conclusions concerning the genesis of the veins and the source from which the gold and other constituents were derived, must be based, in the region under consideration, principally on theoretic considerations and for that reason can not be very satisfactory.

In preparing this report the advisability of testing the rocks of the various auriferous areas by assaying a large number of specimens for gold was considered. It was seen, however, owing to certain conditions, that the results of work of this character would have a very slight value. In the majority of cases, should gold have been detected, it would have been impossible to have determined whether it was original in the rocks or had been deposited there contemporaneously with the vein deposits. In some of the schists and gneisses of the region, it is difficult, even microscopically, to discriminate between original and secondary silica, and numerous tiny veinlets of quartz are of common occurrence. The chances also of selecting specimens from an area slightly impregnated with gold, due either to the presence of a regularly impregnated deposit, or resulting from the close proximity of an undetected auriferous vein, would be great in a region where hot silicated waters have played an important role.

There are certain features of the gold veins in Georgia that point to the deposition of the ores from heated waters coming from great depths. The veins, if of Paleozoic or pre-Paleozoic age, had their roots deep seated when formed. At the Creighton mine in Cherokee county, underground works have shown the continuity of the vein along a steep incline for nearly a thousand feet below the present surface. The Piedmont Plateau, within whose area this and the majority of the deposits are found, has twice been reduced to a condition of base-level subsequent to Paleozoic times—once during the Cretaceous and once during the Tertiary period. Prof. Wm. Davis, speaking of the Piedmont Plateau in Virginia, says "The height to which the rock masses once rose above the present surface is reasonably estimated as at least one mile; it may have been two or three."[1] The extensive formations of sands and clays whose material was derived from the rocks of the Piedmont region that are found south of the Fall Line in Georgia add their testimony to the great erosion of the plateau in past ages.

There are also minerals in some of the veins associated with the gold and gangue material that suggest an origin from hot, very deeply circulating waters if not from the waters of cooling rock magmas.

---

1. Physical Geography, Davis, p. 189.

Several students of ore deposits have been inclined in re-cent years to refer the origin of gold in quartz veins at a number of localities to the agency of heated waters passing off from consolidating and cooling granite and grano-diorite magmas. Mr. A. C. Spencer, at the Washington meeting of the American Institute of Mining Engineers in May, 1905, called attention to the fact that these escaping solutions would carry with them a large part of the highly soluble constituents of the original magma, such as chlorides, fluorides, carbonates, sulphates, etc., some of which if present would increase the dissolving power of the water with respect to silica, the metallic sulphides and gold.

Mr. L. C. Graton, in a paper entitled "Reconnaisance of Some of the Gold and Tin Deposits of the Southern Appalachians" published as bulletin 293 of the U. S. Geological Survey, in the case of a portion of the gold deposits of the Carolinas, is inclined to refer the source of the solutions carrying the gold to granitic magmas. Mr. Graton thinks that the veins in that area were not formed in pre-existing open spaces, but that the pressure which the solutions were under enabled them to force the rocks apart along planes of weakness.

Dr. Becker, in his studies of the Southern Appalachian deposits arrived at the conclusion that the fissure veins, as a rule, were formed in interstitial spaces opened by normal faulting. At the Creighton mine in Cherokee county, the occurrence of much of the ore in beautifully alternating bands of quartz and iron pyrite would seem to argue in favor of the deposition from solution of that deposit in a pre-existing cavity. It has been recognized that a banded structure may, in some cases, be present in replacement deposits, but there seem to be no features at this locality that specially point to a replacement deposit. Plate II, Fig. 2, gives a view of some specimens of this banded ore.

Dr. Becker, in the paper that has been several times referred to, does not go into an extended discussion as to the probable source from which the gold in the veins was derived.

Granites, or granite-gneisses, are of common occurrence in the neighborhood of many of the Georgia deposits. This fact, together with the reasons given above for believing that the solutions came from great depths, might well make a theory referring the source of the gold to the magma of these rocks a tenable one. Unfortunately, in the present stage of knowledge concerning the deposition of ore deposits from a deep seated source, little can be done beyond suggesting this origin as a possible one. In this connection, it is of interest to note the occurrence, at certain localities, of small stringers or lenses of quartz in gneiss or gneissic mica schist that grade into bodies that seem to represent either interfoliated pegmatite dikes or segregated portions of an originally igneous rock. An unsuccessful effort was made to secure a photograph of some pieces of ore of this character from Forsyth county showing a mottled appearance due to the intimate mixture of quartz and feldspar. Free gold in fairly coarse particles was present in this material. While the specimens obtained were too much altered to admit of microscopic examination in regard to whether the quartz or feldspar is of secondary origin the general appearance of the material suggests original constituents.

Another interesting association of minerals was noticed at the Standard (formerly Singleton) mine at Dahlonega. Specimens of quartz were obtained showing gold in thin leaves embedded in mica crystals in such a manner as to leave no doubt as to the contemporaneous deposition of the gold and mica.

At the Loud mine in White county, many beautiful specimens of crystallized gold have been obtained that would seem to indicate some differences at that locality in the conditions that normally prevailed at the time of vein deposition. Plate II, Fig. 1, gives a photographic reproduction of a cluster of these crystals associated with crystals of quartz. That this specimen must have formed under conditions of little pressure and in an open space seems certain. On an adjoining property to the Loud, and along the strike of the auriferous belt, considerable masses of vein quartz were noticed in a prospect pit that showed a marked tendency to crystalization and suggested a comb structure.

In the gold producing region of the State, it is very difficult at many localities among the highly metamorphosed schists and gneisses, to discriminate between rocks of originally igneous and others of possibly sedimentary origin and to refer altered igneous rocks to their parent magmas. This uncertainty regarding the origin of the rocks increases the difficulty of arriving at definite conclusions concerning the source of the gold.

A number of specimens of rocks were collected in the course of the field work of this report from different auriferous areas. The more typical ones were very fully analyzed by Dr. Edgar Everhart, Chemist of the Geological Survey. The results of these analyses, together with a brief description of the microscopic characteristics are to be found at the close of this chapter. Very detailed work, involving probably the study of hundreds of specimens, would be necessary in order to make anything like thorough petrographic investigations in the area. Such a work would be beyond the scope of the present publication, but it is hoped that at a later date more may be done along this line.

In the Dahlonega belt the association of two classes of rocks, one decidedly more basic than the other, is very noticeable. That this association has, in some way, been a factor in the formation of the gold deposits, seems extremely probable, as many of the most productive mines are found either near or at the contact of these two classes of rocks.

The more basic of the two types is represented by rocks containing large amounts of hornblende and they are traceable along the gold belt all the way from Burnt Hickory Ridge in Paulding county into Macon county, North Carolina.[1] At some localities between the points mentioned these rocks are very narrow in lateral extent, but it is doubtful if they are ever entirely wanting. They are present in considerable bodies at Dahlonega, a good exposure occurring in the public square of that town. They all possess gneissoid or schistose structure in varying degrees, at some localities being very fine grained schists showing on fresh fracture to the unaided eye only silky aggregates of hornblende. Owing to the fact that their weathered product is of a deep brownish yellow color and, at some localities, is frequently found in prismatic blocks, it has been universally termed "brickbat" by the miners.

Microscopic study of a number of specimens of these basic rocks shows that they are of igneous origin and certain features common to all of them suggest a close relationship. The ratio to each other, however, of hornblende and quartz (the two most characteristic minerals) varies considerably. The origin of these hornblende schists and gneisses is obscure. Some of them may represent ancient, sheared and altered diorites.

The results, however, of chemical analysis of several specimens from different localities (description of rocks at close of this chapter) indicate more basic rocks than a diorite of average composition.

---

1. The unsheared diabases occuring in well defined dikes throughout the Crystalline area are not included in the basic types here referred to.

In a special map of the geology of the Dahlonega district, by Mr. Arthur Keith of the United States Geological Survey, which is reproduced in this report, these basic rocks occur in the areas designated as the Roan gneiss. By reference to this map it will be seen how many of the mines about Dahlonega, that are there indicated, occur along the contact of the Roan gneiss and the Carolina gneiss in which latter formation are to be found the acid rocks previously mentioned. This map also shows that a few of the mines are at, or near, the contact of granite masses and one or the other of the two formations just mentioned.

The acid rocks of the Dahlonega belt show greater variations both in appearance and mineral composition at different localities than do the rocks of the basic series. Mica schist, quartz-feldspar mica schists and gneisses are the prevailing types. Some of these were doubtless derived from the metamorphism of granites and granite porphyries. Others may represent altered sediments. No positive proof of original sedimentary origin was obtained in any of the specimens studied from the Dahlonega belt. In the Bast cut at Dahlonega a highly siliceous rock is spoken of in the notes on that mine as having a probable sedimentary origin. In a region, however, that has been subjected to repeated orogenic movements and profound dynamic disturbances, allowances must be made for changes in chemical composition. Cases have been observed by the writer in the Crystalline area of Georgia where very siliceous rocks can be seen from field evidence to have been derived from rocks of more normal type, by alteration and silicification alon a shear zone. Several analyses. given at the end of this chapter, of mica schists from different points along the Dahlonega belt fail to throw much light on the question of their probable origin. In the particular cases given, excepting the schist of the Bast and Findley cuts at Dahlonega,.

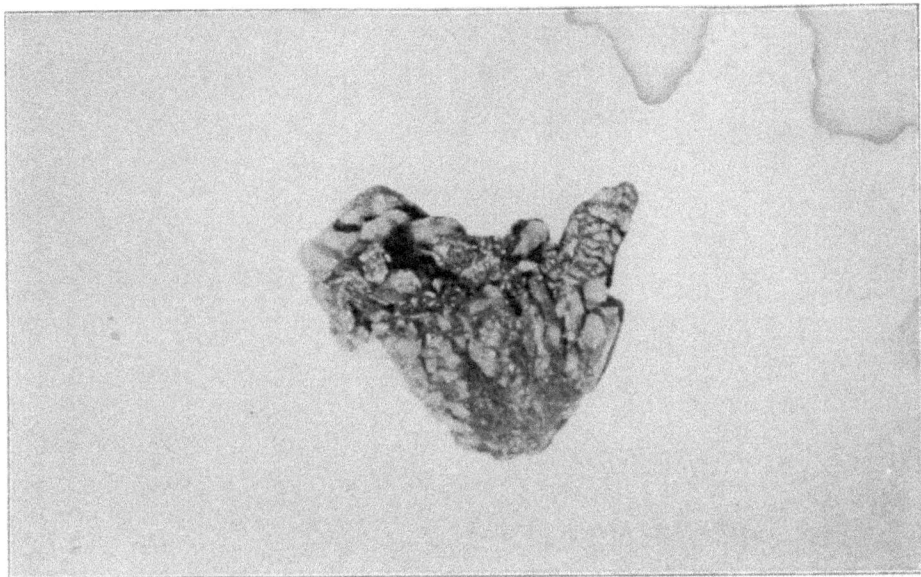

FIG. 1.—GOLD CRYSTALS ASSOCIATED WITH QUARTZ CRYSTALS, LOUD GOLD MINE,
WHITE COUNTY, GEORGIA

FIG. 2.—GOLD ORE FROM THE CREIGHTON GOLD MINE, CHEROKEE COUNTY, GEORGIA,
SHOWING BANDED STRUCTURE

if of sedimentary origin, they must have been formed from sediments of rather heterogeneous character yielding rocks analogous in chemical composition to certain igneous rocks.

In the McDuffie belt the rocks studied show evidence of hav-.ing been derived from rather basic porphyritic granites or from grano-diorites. The marked association of acid and basic rocks, characteristic of the Dahlonega belt, is absent in this gold belt. Indeed, with the exception of the Dahlonega belt, this marked association of two distinct classes of rocks was not observed as a characteristic feature of the auriferous deposits. Areas of dark reddish saprolites occur about a fourth of a mile to the northwest of the Carroll county belt in the vicinity of Villa Rica and doubtless represent a rock rich in some ferro-magnesian mineral. Basic rocks in greater or less amounts are not infrequent in the vicinity of many of the gold bearing veins at various localities throughout the State, but, excepting the Dahlonega belt, the association referred to can not be considered as characteristic. The larger portion of the auriferous veins occur in acidic schists and gneisses and their origin does not appear to have been dependent on the presence of more basic rocks. The association of auriferous deposits at certain localities with pegmatite dikes, or bodies of a closely similar nature, has already been mentioned.

Reference to the map accompanying the text will show that the distribution of the major portion of the auriferous deposits is in relatively narrow parallel belts having an almost identical trend, which is northeast and southwest, the prevailing trend of the formations of the Crystalline area. The areas of the various isolated deposits also show more or less elongation in the same directions as the major belts. The McDuffie belt has a trend differing slightly from that of the other belts. The distribution of the deposits would appear to be due to either one of two causes to be men-

tioned below, or may possibly have resulted from the combined effect of these causes. More detailed study of the
structural geology of the Crystalline area in the region
under discussion may show that the gold belts are located
along zones parallel with the prevailing trend of the formations, in which unusual shearing and metamorphism has taken
place, or along belts in which the association of different
rocks was favorable to the formation of auriferous quartz
veins, or both of these conditions may have been operative
in producing the deposits. In the case of one of the isolated
auriferous localities of Lincoln county, viz., the Seminole
mine, there seems to be little question that the presence of a
shear zone determined the location of the deposit.

Summing up the more important geological facts concerning the deposits, the following is to be noted: 1. By
far the larger portion of the auriferous areas occurs in narrow
parallel belts having the prevailing trend of the formations
of the Crystalline area. 2. In the Dahlonega belt, the most
persistent of the auriferous belts, the association of two
classes of rocks, one much more basic than the other, is
characteristic, and while auriferous quartz veins occur in both
varieties of rock, the contact of the two seems to have determined the location of many of the more important deposits.
3. In belts other than the Dahlonega belt, while basic rocks
are not infrequent, their presence does not seem to have
been an important factor in the location of the deposits, the
greater portion of the auriferous veins being found in acid
schists and gneisses. 4. Bodies of granite, or rocks that there
is reason to believe were derived from the metamorphism
of granite or granite porphyries, are of frequent occurrence in
the gold belts, either immediately contiguous to the auriferous
deposits or sufficiently near to have exercised some influence
on their genesis. 5. At certain localities auriferous veins

are closely associated with and, in some cases, appear to grade into pegmatite dikes or bodies of similar mineral composition. 6. The occurrence and association of certain minerals in the auriferous veins and the character of the deposits indicate a deep seated origin for the ore bodies.

As stated in another part of the chapter, granite magmas may have been the source of the gold of a portion of the auriferous deposits. Taking into consideration the facts above enumerated, it appears probable that granite, or rocks derived from granites or grano-diorites or porphyritic facies of these rocks were at many localities the source of the gold in the quartz veins. It also seems reasonably certain that where basic rocks, at present containing large amounts of hornblende, were associated with more acid rocks in the gold belts the contact of the two classes was a factor in determining the location of unusually important auriferous deposits.

A consideration of the more notable occurrences of gold in various sections of the globe will show the difficulty of formulating any general hypothesis as to the source of gold and the factors controlling its concentration in auriferous veins. Mr. T. A. Rickard, in a paper on the geological distribution of gold, has shown the occurrence of important auriferous deposits in various geological formations from the Archaean to the Quaternary and in a variety of different kinds of rocks.[1] A number of painstaking and laborious investigations have been prosecuted at different times by different geologists in testing the rocks in the vicinity of auriferous deposits for gold contents. The results of work of this class have not, as a rule, afforded satisfactory data from which to draw conclusions. In these investigations of rocks in re-

---

1. Rickard, T. A., The Geological Distribution of Gold. Amer. Mining Congress Papers, Vol. IX, 1906, pp. 105-114.

gard to the probable source of gold in the adjacent vein and other deposits, a consideration of the uncertainty of forming any reliable estimate of the probable amount of gold that may occur widely, and possibly irregularly, distributed through a large area of rock has not apparently been given sufficient weight. The amount of concentration that might take place in veins through natural causes acting during long periods of time and over wide areas of a rock containing gold that could not be satisfactorily tested may be considerable. If a miner can, with the aid of machinery, concentrate the gold from ore containing two-tenths of an ounce per ton sufficiently rapidly to make it a profitable enterprise, it may be conceived that natural causes, acting under the conditions mentioned, can, from rocks containing small amounts of gold, concentrate into a quartz vein sufficient quantities of the metal to form a commercially valuable deposit. Where rocks have been tested for auriferous contents the amounts taken from any single point have not probably been sufficiently large to make the determinations of much value as indicating even the probable quantities of gold present. In the case of the elaborate tests made in the Australian gold fields by Dr. John R. Don, it is stated that, in the majority of cases, the samples taken weighed seven or eight pounds.[1] While the presence or absence of gold would probably be proven by tests of this character an approximate knowledge of the average amount in a considerable area of rock, the most important data from which to draw conclusions, would not be obtained. In a rock containing a hundredth of an ounce of gold per ton, if the metal was evenly distributed, for each ton of rock the quantity mentioned would have to be divided into as many as two hundred and fifty particles to insure one particle of gold to each eight

---

1. Amer. Inst. Mining Engineers, Vol. XXVII, 1897, p. 565.

pounds of rock. In actual occurrence the gold is not probably thus uniformly distributed. Unless, therefore, considerable bulks of rock were tested, or many small samples from numerous localities mixed together before the determinations were made, the data obtained would be largely the results of chance. The difficulty also, as was pointed out by Posepney in a discussion of Becker's studies of the Comstock lode,[1] of determining the original or secondary character of ore forming material in rocks adds to the uncertainty of drawing conclusions from work of this class.

The knowledge at present available concerning auriferous deposits shows that gold in small quantities is widely distributed through much of the earth's crust and that at many localities, notably, in regions of igneous and crystalline rocks, it is found concentrated in commercial quantities in mineral veins. It is known that concentration in veins is favored by dynamic disturbances and attendant metamorphism, but sufficient data are not yet available from which to formulate generalizations in regard to either the sources or processes of concentration that can be applied very satisfactorily to specific occurrences.

DESCRIPTION OF ROCKS

The general characteristics of the several classes of rocks associated with the Georgia gold deposits have already been mentioned in the preceding pages of this chapter. More detailed descriptions are given below of some of the principal features of a number of specimens of rocks selected from various auriferous areas. Owing to the uncertainty of determining the origin of a part of these highly metamorphosed crystallines, any classification from a genetic standpoint is, in some cases, open to doubt.

---

1. Genesis of Ore Deposits, Prof. Franz Posepney, p. 86.

The chemical analyses accompanying the descriptions of most of the specimens were made by Dr. Edgar Everhart, Chemist of the Geological Survey. These analyses will be of special interest to petrographers as, with the exception of a number of analyses of granites and granite-gneisses by Dr. Thos. L. Watson, which are embodied in Bulletin 9-A of this Survey, very few complete analyses of rocks from the Crystalline area of Georgia have ever been published.

QUARTZ-ALBITE PORPHYRY, NEAR THE SEMINOLE MINE, LINCOLN COUNTY.—This rock occurs in the vicinity of the Seminole Gold and Copper mine in the western part of Lincoln county. The specimen described was collected from the ravine of a small branch at a point a hundred yards or more to the northwest of the Magruder vein, the northwesternmost of the three principal veins of the Seminole mine. This rock occurs over a considerable area in the vicinity of the mine and it is shown in the chapter on description of individual properties that the ore bodies are located in a highly quartzose sericite schist derived from the alteration and silicification in a shear zone of the rock under discussion.

Megascopically, the rock is medium fine grained and shows to the unaided eye a more or less schistose structure. A conspicuous feature is the presence of numerous spherical, opalescent quartz anhedra, some of which show imperfectly developed pyramidal crystal faces. The residual material of the weathered rock can be identified over large areas by the presence of these quartzes, strewn unaltered over the surface. By this means its saprolite can be traced for a considerable distance to the west of the ravine above mentioned. Southeast of the Seminole mine for some distance the surface residual material affords little clue to the nature of the underlying rock, but something like an eighth or quarter of a mile in that direction the characteristic quartzes are to be found in the soil over an extensive area.

A thin section of the specimen shows an igneous rock possessing the structure of a granite porphyry. Numerous quartz anhedra together with phenocrysts of albite feldspar and flakes of biotite in smaller amounts, are scattered through a fine grained ground-mass of quartz and plagioclase feldspar in intimate intergrowth. A portion of the ground-mass has a rather course spherulitic structure, such as is sometimes found in small dikes or in larger bodies of certain igneous rocks near their peripheries. Occasional areas of the ground-mass are made up principally of feldspar microlites showing a confused, roughly parallel arrangement apparently due to flowage. Magnetite is present as an accessory mineral and crystals of pyrite that are noticeable in the hand specimen are seen in the slide. Short, stout crystals of apatite of relatively large dimensions occur in sparing amounts. The quartz anhedra contain fluid and gas inclusions and frequently show round and pear shaped inclusions of the ground-mass. They also not infrequently show resorption or corrosion phenomena. The albite phenocrysts occur principally in short, stout forms and, like the quartz, many of them appear to have been rounded by resorption. Much of the biotite occurs as flakes and shreds wrapping around the feldspar phenocrysts and the quartz anhedra. and it is also to be found around the spherulites in the ground-mass. A considerable portion of it has altered to weakly pleochroic chlorite.

The analysis of this interesting rock is given below. Over six per cent. of soda is present with a little less than one per cent. of lime and still smaller amounts of potash.

Silica 68.07
Alumina 15.07
Ferric oxide 1.13
Ferrous oxide 3.42
Magnesia 2.27
Lime . ... .73
Soda 6.06
Potash . ... .29
Loss on ignition 1.25
Moisture at 100° C. .... .01
Carbon dioxide .... .00
Titanium dioxide .... .37
Phosphorus pentoxide .... trace
Sulphur 1.14
Manganous oxide .... .31

Total..100.12

SHEARED AND ALTERED QUARTZ-ALBITE PORPHYRY.—This rock was collected at the same locality as the specimen above described, but at a point lower down the ravine and about thirty yards from the extension of the Magruder vein previously mentioned as being the northwesternmost of the veins of the Seminole mine. The rock here is an altered phase of the one higher up the ravine. It is much more schistose in structure and many of the quartz anhedra and the feldspar phenocrysts are no longer visible. Close inspection of a hand specimen will, however, reveal a few of the characteristic quartzes. Under the microscope it is seen that this rock is a sheared phase of the one first described. The quartz anhedra can still be seen, but some of them are badly crushed and granulated. White mica has developed as a new mineral, many of the areas originally occupied by feldspar phenocrysts being now almost completely filled with this material. The presence of some orthoclase shows a slight variation in the original character of the two rocks.

The chemical analysis here follows. The percentage of silica is practically the same as in the preceding specimen.

though as will be seen from the description of the next specimen taken from within the zone of the ore bodies, some silicification has there taken place.

Silica	68.13
Alumina	17.49
Ferric oxide	1.19
Ferrous oxide	1.38
Magnesia	1.00
Lime	.75
Soda	3.50
Potash	3.00
Loss on ignition	2.06
Moisture at 100° C.	.03
Carbon dioxide	.00
Titanium dioxide	.74
Phosphorus pentoxide	trace
Sulphur	.22
Manganous oxide	.30
Total	99.79

QUARTZ-SERICITE SCHIST, SEMINOLE MINE, LINCOLN COUNTY. —This specimen was collected a short distance to the southeast of the last described rock and within the zone of the ore bodies of the Seminole mine. It represents a highly sheared and altered phase of the quartz-albite porphyry accompanied with a certain amount of silicification. This rock is a very fine grained, light colored, friable schist which is seen in thin sections to be composed principally of quartz grains and white mica. Remnants of the quartz anhedra observed in the other rocks collected from the locality are still discernable and suggestions of original feldspar phenocrysts are noticed in the areas of the white mica, although most of this mineral occurs as narrow bands or layers between the quartz grains. A little biotite is present. The chemical analysis of this rock is given below. It shows that the percentage of silica has increased from sixty-eight per cent. in the specimens previously described to over seventy-seven in this one.

Silica 77.12
Alumina 12.65
Ferric oxide 2.60
Ferrous oxide .... .73
Magnesia 1.08
Lime . ... trace
Soda 2.07
Potash . ... .21
Loss on ignition 2.64
Moisture at 100° C. .... .10
Carbon dioxide .... .00
Titanium dioxide .... .28
Phosphorus pentoxide .... .00
Sulphur . ... .08
Manganous oxide .... .31

Total.. 99.87

GRANITE PORPHYRY, ONE MILE EAST OF THE SEMINOLE MINE, LINCOLN COUNTY.—Southeast of the Seminole mine, commencing at a point about an eighth of a mile from the mine, the characteristic quartz anhedra found in the rocks previously described occur in the soil over a wide area. Approximately a mile east of the mine, there occurs a limited outcrop of a badly weathered light colored porphyritic rock. It contains numerous quartz anhedra similar in character to those in the rocks to the west.

In thin section, this rock is seen to be quite similar in appearance to the quartz-albite porphyry at the Seminole mine. The specimen sectioned is too badly altered to permit of satisfactory study. It contains, however, numerous anhedral quartzes that are identical in character with those of the other rocks studied. The rock is a porphyritic with a fine grained ground-mass, but no spherulitic structures are noticeable as in the case of the quartz-albite porphyry. Biotite is also absent, but as considerable epidote is present some of the former mineral may have been one of the original constituents.

The chemical analysis here follows:

Silica .	69.30
Alumina .	15.91
Ferric oxide	3.20
Ferrous oxide	.18
Magnesia .	.21
Lime .	5.92
Soda .	3.35
Potash .	.14
Loss on ignition	1.80
Moisture at 100° C.	.52
Carbon dioxide	.00
Titanium dioxide	.48
Phosphorus pentoxide	.00
Sulphur .	.03
Manganous oxide	.11
Total	101.15

GNEISS, BELL SHAFT, COLUMBIA MINE, McDUFFIE COUNTY.— This specimen was collected from the dump of the Bell shaft a few hundred yards to the north of the main Columbia mine. The rock is a light colored, medium grained one with a gneissoid structure and appearing to the unaided eye to contain very little quartz.

In thin section it is seen to be composed principally of feldspar and quartz with smaller amounts of biotite and chlorite and some secondary calcite. Magnetite is present as the chief accessory mineral. The larger portion of the feldspar occurs as phenocrysts of a plagioclase that are badly clouded with flakes of white mica. Microperthite is present in smaller amounts as irregular shaped masses. In some cases it shows as a secondary growth fringing the larger feldspar phenocrysts. The most of the quartz occurs as small grains, but larger pieces here and there, surrounded by smaller fragments, suggest the presence of original large quartz anhedra. Some secondary quartz is present as is shown by the occurrence of tiny veinlets of this mineral cementing broken feldspar crystals. Biotite, in small amounts, is present as shreds and patches and is largely altered to chlorite. Micrographic intergrowths of quartz and feldspar are noticeable and though the rock has been considerably crushed and altered it is plainly seen that it has been derived from an igneous rock of a granitic type. The amount of magnetite present is rather greater than is usual in Georgia granites.

The chemical analysis of this rock is given below.

Silica .	63.82
Alumina .	15.94
Ferric oxide	2.91
Ferrous oxide	3.87
Magnesia .	1.20
Lime .	2.91
Soda .	2.54
Potash .	1.80
Loss on ignition	3.70
Moisture at 100° C.	.06
Carbon dioxide	.21
Titanium dioxide	.55
Phosphorus pentoxide	.04
Sulphur .	trace
Manganous oxide	.18
Total	99.73

MICA SCHIST, COLUMBIA MINE, McDUFFIE COUNTY.—The specimen here described came from near the bottom of the working shaft of the Columbia mine. It is one of the most characteristic rocks of the McDuffie gold belt. Megascopically it is a very fine grained, dark gray schist. The only component mineral that can be identified with the unaided eye is sericitic mica in tiny glistening flakes.

In thin section, the schistose structure is even more apparent than in the hand specimen. Innumerable flakes and shreds of white mica, arranged in parallel bands or layers, make up a large percentage of the constituents. The spaces between the layers of mica are occupied principally by grains of quartz, and some plagicolase feldspar is probably present as is indicated by the analysis. A noticeable amount of chlorite occurs in the form of shreds and irregular shaped patches, and small grains of calcite are present in considerable amounts. A number of crystals and grains of iron pyrite, partially altered to iron oxide, are scattered irregularly among the other minerals.

The similarity in chemical composition, as is shown by the analysis, of this rock and the gneiss at the Bell shaft, together with the close proximity of the latter rock, suggests that the mica schist is a highly sheared phase of the gneiss.

The chemical analysis of the schist is given below.

Silica .	62.45
Alumina .	15.91
Ferric oxide	2.07
Ferrous oxide	3.80
Magnesia .	1.76
Lime .	3.70
Soda .	2.33
Potash .	.40
Loss on ignition	5.41
Moisture at 100° C.	.03
Carbon dioxide	.50
Titanium dioxide	.82
Phosphorus pentoxide	.13
Sulphur .	.22
Manganous oxide	.37
Total	99.90

MICA SCHIST, PARKS MINE, McDUFFIE COUNTY.—The specimen was collected from the dump of the Parks mine which is about a mile northeast of the Columbia mine.

This rock is quite similar in appearance in the hand specimen to the mica schist at the Columbia mine. As will be seen from the analysis it is slightly more basic in character than the latter rock. A number of specimens at the Parks mine were noticed that showed corrugations on the planes of·schist-osity.

In thin section, it is seen to be somewhat different from the rock at the Columbia mine. Mica is less abundant and a good deal of epidote is present. Chlorite occurs as in the rock from the last named locality and the difference between the two specimens is not great enough to cause them to be considered as belonging to separate series.

The chemical analysis here follows:

Silica	61.25
Alumina	17.18
Ferric oxide	4.49
Ferrous oxide	2.54
Magnesia	2.21
Lime	3.78
Soda	2.50
Potash	1.60
Loss on ignition	3.03
Moisture at 100° C.	.01
Carbon dioxide	.00
Titanium dioxide	.92
Phosphorus pentoxide	.02
Sulphur	.24
Manganous oxide	.27
Total	100.04

MICA SCHIST, BELL SHAFT, COLUMBIA MINE, McDUFFIE COUNTY.—The specimen was collected from the dump of the Bell shaft at the Columbia mine. According to Mr. W. H. Fluker, general manager, it and the gneiss previously described from that locality are both encountered in the underground works on the Bell vein. This schist is seen from a thin section to be so closely related to the mica schist at the Columbia mine that it is unnecessary to describe it in detail.

WALL ROCK, EGYPT SHAFT, POTER MINE, McDUFFIE COUNTY. —This rock was collected from the dump of the Poter mine which is situated about two and a half miles northeast of the

Parks mine. The only material obtainable was badly weathered. Megascopically it suggests a rock of a granitic type that has been considerably crushed and sheared. Phenocrysts of partially kaolinized feldspar, which is the only mineral determinable with the unaided eye, give it a speckled appearance.

In thin section, it is seen that the rock is an igneous one of a granitic type that has been badly crushed and altered. Phenocrysts of partially altered plagioclase feldspar are the most prominent constituent. Microperthite, in smaller irregular shaped masses, is sparingly present and small grains of quartz are numerous. Micrographic intergrowths of quartz and feldspar are common. Biotite, most of it altered to chlorite, occurs in small quantities and a large amount of secondary calcite is present. While the specimen examined is too altered to permit of positive identification, it appears to be very similar in character to the gneiss at the Bell shaft of the Columbia mine.

WALL ROCK, MORGAN MINE, OGLETHORPE COUNTY.—The specimen was collected from the hundred and ten foot level of the Morgan mine. Megascopically, it is a fine grained schistose rock, dark gray in color with a decided greenish tone. Under the microscope it is seen that it is an igneous rock possessing originally a granitic structure, but now badly sheared and altered. Remnants of original phenocrysts of a plagioclase feldspar are readily discernable, though they are badly clouded with scales of white mica or kaolin. Quartz is an important constituent and frequently shows micrographic intergrowths with feldspar. Sericitic mica is present in large amounts, and a good deal of chlorite occurs in the form of shreds and irregular shaped patches. Magnetite is noticeable as an accessory mineral and some secondary calcite is present and a little epidote. This rock probably represents a sheared and altered grano-diorite.

The chemical analysis is given below.

Silica .	58.71
Alumina .	14.96
Ferric oxide	1.70
Ferrous oxide	4.59
Magnesia .	3.19
Lime .	4.90
Soda .	3.24
Potash .	.94
Loss on ignition	3.89
Moisture at 100° C.	.05
Carbon dioxide	2.20
Titanium dioxide	.83
Phosphorus pentoxide	.04
Sulphur	.08
Manganous oxide	.28
Total	99.60

HORNBLENDE-ZOISITE SCHIST, CREIGHTON MINE,, CHEROKEE COUNTY.—This specimen was collected from the dump of the Creighton mine. Megascopically, the rock is a fine grained, dark colored schist, showing at some points the coarser banding of a gneiss rather than a schist.

In thin section it is seen that the most predominant mineral is common green hornblende occurring as fibers and elongated ragged prisms arranged parallel with the planes of schistosity. Some of it shows imperfectly developed cleavage, but it is most probably a secondary hornblende. Zoisite and quartz are the two next most important constituents. The quartz occurs as grains, principally between the hornblende fibers. It is, in part at least, secondary in origin. The zoisite is partly in the form of grains and irregular shaped, elongated pieces and also occurs as larger, more or less rounded or short columnar, masses. In the latter form it frequently contains included grains of quartz, and, in the central portions, small inclusions such as are common in hornblende crystals. The hornblende fibers, and other minerals are frequently more or less wrapped around these larger pieces of zoisite. Some grains of a plagioclase feldspar are scattered throughout the slide.

The origin of a rock so highly metamorphosed and recrystallized must necessarily be open to doubt. It may represent a sheared and altered diorite.

The chemical analysis is given below.

Silica	51.91
Alumina	18.02
Ferric oxide	2.38
Ferrous oxide	6.42
Magnesia	5.69
Lime	11.00
Soda	.43
Potash	2.62
Loss on ignition	.83
Moisture at 100° C.	.00
Carbon dioxide	.00
Titanium dioxide	.71
Phosphorus pentoxide	.02
Sulphur	trace
Manganous oxide	.27
Total	100.30

QUARTZ-FELDSPAR-MICA SCHIST, NEAR THE CREIGHTON MINE, CHEROKEE COUNTY.—The specimen was collected a short distance to the southwest of the Creighton mine. Megascopically the rock is a gray schistose one showing to the unaided eye glistening scales of mica and thin layers of some different substances of a light color.

Under the microscope it is seen to be made up of numerous flakes of both biotite and white mica and grains of quartz and a plagioclase feldspar. A number of small crystals of garnets are also present, and a little magnetite and some iron pyrite occur. It probably represents a highly sheared and crushed granite, though positive proof of such an origin is lacking.

The chemical analysis is given below:

Silica	63.46
Alumina	15.97
Ferric oxide	2.37
Ferrous oxide	4.59
Magnesia	1.89
Lime	2.76
Soda	1.27
Potash	3.01
Loss on ignition	1.13
Moisture at 100° C.	.00
Carbon dioxide	.00
Titanium dioxide	1.66
Phosphorus pentoxide	.04
Sulphur	1.46
Manganous oxide	.31
Total	99.92

AMPHIBOLITE, BURNT HICKORY RIDGE, PAULDING COUNTY.—
The specimen was collected from the Huntsville public road
on Burnt Hickory Ridge a mile or so southwest of the Dunaway gold mine. Megascopically it is a very fine grained, dark
colored schist, none of whose component minerals can be satisfactorily identified with the unaided eye.

In thin section, it is seen to be similar in several respects
to the hornblende schist, previously described, from the
Creighton mine. It contains, however, a considerably larger
prcentage of hornblende than the Creighton mine rock and
less quartz. Zoisite and epidote are present in the form of
grains and small, irregular shaped masses. Much of the
hornblende occurs as slender prisms and fibers, but there are
some short stout prisms containing numerous inclusions that
appear to represent an older generation of crystals than the
balance of this material.

The chemical analysis is given below. It is seen from this that the rock is more basic in character than the hornblende schist at the Creighton mine, but the mineral composition and general similarity of the two rocks suggest a closely related origin.

Silica	48.67
Alumina	13.18
Ferric oxide	10.18
Ferrous oxide	.78
Magnesia	7.82
Lime	10.68
Soda	2.71
Potash	.03
Loss on ignition	.66
Moisture at 100° C.	.10
Carbon dioxide	.00
Titanium dioxide	1.42
Phosphorus pentoxide	.03
Sulphur	.02
Manganous oxide	.04
Total	99.32

OLIVINE DIABASE, CREIGHTON MINE, CHEROKEE COUNTY.—
The specimen was collected from a small dike that is exposed in shaft No. 3½ of the Creighton mine. Where this dike is exposed in the shaft it intersects the Franklin vein, the principal auriferous quartz vein of the mine. Megascopically, it is a dark rock showing the dense, fine grained structure characteristic of diabases occurring as small dikes.

In thin section it is seen to be a typical olivine diabase and merits no detailed description. Olivine is quite abundant, and occurs both as large grains and poorly developed crystals, and also as smaller well developed crystals.

The chemical analysis here follows:

Silica .	45.23
Alumina .	15.73
Ferric oxide	.06
Ferrous oxide	11.47
Magnesia .	12.29
Lime .	8.33
Soda .	4.72
Potash .	.07
Loss on ignition	.70
Moisture at 100° C.	.39
Carbon dioxide	.00
Titanium dioxide	1.10
Phosphorus pentoxide	.02
Sulphur .	.08
Manganous oxide	trace
Total	**100.19**

MICA SCHIST, CHILDS MINE, NACOOCHEE VALLEY, WHITE COUNTY.—The specimen was collected from the dump of a prospect shaft a short distance from the milling plant of the Childs mine on the Dahlonega gold belt. Megascopically, it is a gray, fine grained schistose rock showing glistening scales of mica.

In thin section it is seen to be a typical quartz-mica schist. Numerous plates of both biotite and white mica occur in roughly parallel layers, while the spaces between the mica scales are occupied principally by interlocking quartz grains. Some chlorite is present and a few grains of unstriated feldspar are mixed with the quartz. Small, short prisms of tourmaline occur in sparing amounts scattered irregularly through the slide. Magnetite is present as an accessory mineral and a small amount of pyrite also occurs. The rock is so thoroughly metamorphosed and recrystallized that, in the absence of confirmatory field evidence, no trustworthy conclusions can be drawn as to its probable origin.

The chemical analysis is given below:

Silica 70.99
Alumina 14.76
Ferric oxide 1.33
Ferrous oxide 3.08
Magnesia 1.75
Lime 1.13
Soda 1.87
Potash 2.69
Loss on ignition 1.32
Moisture at 100° C. .... .09
Carbon dioxide .... .00
Titanium dioxide .... .60
Phosphorus pentoxide .... .04
Sulphur . ... .04
Manganous oxide .... .46

Total.....................................100.15

SHEARED GRANITE PORPHYRY, BARLOW MINE, LUMPKIN COUNTY.—The specimen was collected from an exposure in the large Barlow cut near Dahlonega. Megascopically, the rock is a gray schistose one of rather fine grain, having a very distinct evenly laminated structure. Numerous, bluish opalescent, roughly spherical bodies of quartz of larger size than the average of the mineral constituents are noticeable in the hand specimen. On weathered surfaces these quartzes are one of the most characteristic features of the rock.

In thin section the laminated structure is seen to be due to bands, or layers, of mica alternating with the other constituents. The mica bands are composed of both biotite and muscovite. The spaces between the layers of mica are occupied by a fine grained aggregate of quartz and feldspar grains through which are distributed the roughly spherical bodies of quartz previously mentioned and also areas of large flakes of muscovite usually accompanied with secondary quartz and calcite. Some of these areas occupied by the minerals last mentioned show remnants of original feldspar phenocrysts.

In the fine grained aggregate of quartz and feldspar, quartz is much more abundant than feldspar. The roughly spherical quartz bodies are in many cases badly granulated and considerably elongated in the schistosity planes. Some that have escaped much crushing show inclusions of the fine grained aggregate of quartz and feldspar. There seems little doubt that this rock represents a highly sheared phase of a granite porphyry, the more or less flattened bodies of quartz just described having been spherical quartz anhedra in a porphyritic rock.

The chemical analysis is given below:

Silica	67.36
Alumina	15.06
Ferric oxide	.80
Ferrous oxide	2.75
Magnesia	.96
Lime	2.49
Soda	4.49
Potash	2.02
Loss on ignition	2.23
Moisture at 100° C.	.00
Carbon dioxide	.38
Titanium dioxide	.61
Phosphorus pentoxide	.00
Sulphur	.03
Manganous oxide	.11
Total	99.29

HORNBLENDE SCHIST, DAHLONEGA.—The specimen was collected near the base of Findley Ridge where an excavation had been made in grading the Middle Gainesville public road. Megascopically, the rock is a medium to fine grained schist of a dark appearance but flecked with light colored specks. It is a typical specimen of the rock whose partially weathered product is locally known as "brickbat."

In thin section it is seen that hornblende is the most predominant mineral. Quartz and calcite are the next most abundant constituents. A small amount of plagioclase feldspar is present and pyrite with greater amounts of magnetite or ilmenite occur in rather large quantities for accessory minerals. A small amount of epidote is also present.

The hornblende is of the common green variety and occurs as fibers, and elongated, ragged prisms with poorly developed cleavage. The calcite is generally associated with the hornblende and in some cases it can be seen to be an alteration product of that mineral. The quartz occurs as irregular, interlocking grains. Much of it shows little or no undulous extinction and is probably secondary in origin. A small amount of plagioclase feldspar in the form of grains is mixed with the quartz. The origin of the rock is obscure. It may have been derived from a diorite. It will be noticed, however, that it is a decidedly more basic rock than a diorite of average composition.

The chemical analysis is given below:

Silica .	46.00
Alumina .	15.65
Ferric oxide	.03
Ferrous oxide	10.53
Magnesia .	6.31
Lime .	12.35
Soda .	2.35
Potash .	.03
Loss on ignition	1.40
Moisture at 100° C.	.09
Carbon dioxide	3.66
Titanium dioxide	1.38
Phosphorous pentoxide	.00
Sulphur .	.27
Manganous oxide	.73
Total	100.78

QUARTZ SCHIST, BAST CUT, DAHLONEGA.—The specimen was. collected from an exposure in the southwest end of the Bast. cut on Findley Ridge. Megascopically, the rock is a bluish.

gray, fine grained one showing an imperfectly developed schistose structure. In the specimen here described the texture is of a decided granular, sugary character.

In thin section it is seen that the rock is practically made up of quartz with considerable amounts of accessory magnetite. Some silicate of iron is also probably present as is indicated by the amount of ferrous iron shown in the analysis. The quartz is principally in the form of irregular shaped interlocking grains, but some of the grains are fairly round in outline suggesting an original sedimentary origin. No rims of secondary silica, however, were noticed surrounding any of the grains. The magnetite occurs principally as crystals of varying dimensions scattered rather uniformly throughout the slide. Both microscopic study and the chemical analysis suggest a sedimentary origin for this rock, though positive proof of such origin is lacking.

The chemical analysis here follows:

Silica	88.20
Alumina	.96
Ferric oxide	2.69
Ferrous oxide	7.98
Magnesia	.12
Lime	.00
Soda	.18
Potash	.00
Loss on ignition	.14
Moisture at 100° C.	.00
Carbon dioxide	trace
Titanium dioxide	.05
Phosphorus pentoxide	trace
Sulphur	trace
Manganous oxide	.15
Total	100.47

GARNETIFEROUS QUARTZ SCHIST, FINDLEY MINE, DAHLONEGA. —The specimen was collected from the dump of a shaft in the west cut of the Findley mine. Megascopically, it is a dark gray, quartzitic rock having a distinctly schistose structure.

In thin section it is seen that quartz, in the form of inter-locking grains, is the most abundant constituent. Biotite, largely altered to chlorite, and garnets are next in relative order of abundance. Some white mica is also present and small amounts of pyrite and magnetite. A little calcite also occurs. The garnets, occurring principally as long drawn out, irregular shaped masses, are a conspicuous feature of the rock when examined with the microscope. As in the case of the rock at the Bast cut, a sedimentary origin is suggested for this schist though definite proof to that effect is lacking.

The chemical analysis here follows:

Silica .	73.63
Alumina .	8.93
Ferric oxide	1.02
Ferrous oxide	6.83
Magnesia .	1.72
Lime .	.96
Soda .	1.02
Potash .	2.07
Loss on ignition	.92
Moisture at 100° C.	.00
Carbon dioxide	.00
Titanium dioxide	.55
Phosphorus pentoxide	.00
Sulphur .	.03
Manganous oxide	1.48
Total	99.16

GRANITE, BENNING MINE, DAHLONEGA.—The specimen was collected from the Benning mine on Yahoola Creek. Mega-scopically, the Benning granite is a light colored, rather coarse grained rock containing large flakes of biotite and smaller glistening scales of sericitic mica. In thin section it is seen to be composed of feldspar, quartz, biotite, white mica, some secondary calcite and a little epidote. The feldspar occurs as large grains and poorly developed crystals, and there are also numerous small grains and some small tabular crystals of this mineral. Extinction angles for determining the species were not obtainable in the slide examined. The most of this material is probably albite or an oligoclase high in soda. Numerous plates of white mica have developed in the feldspars. The quartz is in the form of interlocking grains and presents the usual characteristics of the quartz of granitic rocks. The biotite occurs as large plates irregularly distributed through the slide, and a small amount of epidote is usually associated with it. The rock shows distinct evidence of crushing. Some of the larger flakes of biotite are bent and broken and a mortar structure is noticeable at places in the quartz and feldspar grains. The original granular structure, however, has been preserved.

The chemical analysis is given below:

Silica	70.38
Alumina	16.32
Ferric oxide	.50
Ferrous oxide	.75
Magnesia	.60
Lime	3.09
Soda	6.31
Potash	.03
Loss on ignition	1.33
Moisture at 100° C.	.04
Carbon dioxide	.00
Titanium dioxide	.14
Phosphorus pentoxide	trace
Sulphur	trace
Manganous oxide	.08
Total	99.57

SHEARED GRANITE, HAND CUT, CONSOLIDATED MINE, DAHLONEGA.—The specimen was collected from the dump of the main shaft in the Hand cut. It is a fine grained, light colored rock showing a schistose structure. In thin section it is seen to be composed of interlocking grains of quartz and feldspar with parallel layers of white mica. A little chlorite and some calcite are also present and a small amount of epidote occurs. In polarized light the feldspar grains present a blotched appearance showing that they are not of uniform composition. The greater part of the feldspar is probably albite. The field relations between this rock and the Benning granite are not clear, but an original magmatic connection between the two is suggested.

The chemical analysis of this sheared granite is given below:

Silica .	61.66
Alumina .	17.31
Ferric oxide	1.56
Ferrous oxide	5.94
Magnesia .	2.44
Lime .	2.38
Soda .	4.41
Potash .	1.89
Loss on ignition	.53
Moisture at 100° C.	.06
Carbon dioxide	.00
Titanium dioxide	1.47
Phosphorus pentoxide	.00
Sulphur .	trace
Manganous oxide	.18
Total	99.83

MICA SCHIST, HAND CUT, CONSOLIDATED MINE, DAHLONEGA. —The specimen was collected from the dump of the main shaft in the large Hand cut. Megascopically, it is a fine grained, dark gray rock showing a very pronounced schistose structure.

In thin section it is seen to be composed principally of quartz and biotite. The biotite occurs both as sharply defined foils and also as ragged flakes and patches all having their long diameters parallel with the planes of schistosity. The quartz forms a fine grained mosaic, though some of the grains show a distinct elongation parallel with the plates of mica. Very little feldspar is present. Sphene, apatite and magnetite occur as accessory minerals and some pyrite is also present. Much of the sphene is in the form of long ragged masses oriented parallel with the mica. No very trustworthy conclusions can be drawn as to the origin of the rock from the evidence afforded by the section.

SHEARED GRANITE (?), SINGLETON, OR STANDARD, MINE, DAHLONEGA.—The specimen was collected from the dump of the main shaft of the Standard, or Singleton, mine. Megascopically, it is a fine grained, light colored rock showing a sheared structure.

In thin section it is seen to be composed principally of quartz, feldspar and white mica. It resembles in general characteristics the sheared granite at the Hand cut, but the percentage of quartz is greater and much of the feldspar occurs as grains, or phenocrysts of larger dimensions than the remainder of the constituents. It probably represents a sheared granite dike.

SHEARED GRANITE (?), SOUTHWEST END OF SINGLETON CUT, DAHLONEGA.—The specimen was collected from a point near the southwest end of the large Singleton cut. Megascopically, it resembles closely the rock from the main shaft of the Standard mine.

In thin section it is also seen to resemble in general characteristics the rock just mentioned. In addition, however, to the larger grains and phenocrysts of feldspar spoken of in the description of the Standard mine rock it contains some relatively large areas of interlocking quartz grains that may represent original quartz anhedra, larger than the average of the mineral constituents, that have been crushed in shearing.

The chemical analysis here follows:

Silica .	70.50
Alumina .	15.65
Ferric oxide	.44
Ferrous oxide	.67
Magnesia .	.09
Lime .	2.38
Soda .	4.06
Potash .	2.53
Loss on ignition	2.46
Moisture at 100° C.	.01
Carbon dioxide	.20
Titanium dioxide	.18
Phosphorus pentoxide	.10
Sulphur .	trace
Manganous oxide	trace
Total	99.27

METAMORPHOSED SANDSTONE, COHUTTA GOLD MINE, MURRAY COUNTY.—The specimen was collected from the Cohutta Mine on the crest of Cohutta mountain. Megascopically. it is a dark gray, fine grained rock showing a poorly developed schistose structure.

In thin section it is seen to be a sheared and altered arkosic sandstone. The original grains are seen to have been derived from both quartz and feldspars and in some cases an individual grain is composed of both these minerals. The feldspar grains are more or less kaolinized and in the crushing and shearing that has taken place both the quartz and feldspar grains have been somewhat granulated and many of them considerably flattened and elongated. Fine aggregates of quartz and feldspar with chlorite and mica and scales of kaolin occupy the spaces between the grains. A little calcite is also present and some grains of iron oxide. While the specimen described would be classed as an altered sandstone, it is probably a phase of a sheared conglomerate that occurs in the vicinity of the Cohutta mine.

The chemical analysis here follows:

Silica 71.51
Alumina 13.64
Ferric oxide 20
Ferrous oxide 2.51
Magnesia 87
Lime 2.29
Soda 3.23
Potash 2.23
Loss on ignition 1.90
Moisture at 100° C. 04
Carbon dioxide 28
Titanium dioxide 51
Phosphorus pentoxide 09
Sulphur 01
Manganous oxide 09
                                                 _____
Total.. 99.40

CHAPTER V

DESCRIPTIONS OF INDIVIDUAL PROPERTIES

Introductory

The notes on properties, forming the text of this chapter, are given for each gold belt under separate sub-headings. Along those belts passing through more than one county, the different counties are taken up in order of geographical succession from southwest to northeast. The same order is followed in a general way in the description of individual properties, though in the wider portions of the belts it is sometimes deviated from for the sake of convenience. Following the description of properties located on well defined belts there will be found a description of the properties of isolated areas.

Many of the descriptions given are necessarily brief, owing to the fact that no mining operations were in progress at the time of examination. At numbers of important properties, especially where extensive hydraulic operations had been carried on in saprolite material, the veins were found concealed by rain-wash and caving, rendering an examination impossible.

The two conditions just stated also prevented the securing of assay samples at many localities where it would have been desirable to have made assay tests. This deficiency has been partially supplied by quoting the results of assays contained in a former report on a part of the gold deposits of the State published as Bulletin 4-A of the Geological Survey of Georgia. All assays given, with the exception of those quoted from the publication just mentioned, unless otherwise noted, were made by Dr. Edgar Everhart, Chemist of the Geological Survey of Georgia, in the laboratory of the Survey.

THE McDUFFIE BELT

WARREN COUNTY

The southwest end of the McDuffie belt, as defined at present, is found in the northeastern part of Warren county. The portion of the belt lying in this county has not, so far, proved very important from a commercial standpoint. Mining operations have been confined to a limited area immediately west of the McDuffie county line in the vicinity of Cadley. Regular mining has only been conducted at one point, namely, the Warren mine. Gold has been found at several other localities, but no mining operations, other than prospect work have been carried on.

WARREN MINE.—This mine is located in the northeastern part of the county in the neighborhood of Cadley and close to the McDuffie county line. Mining operations were conducted here in the eighties for about two years on an auriferous quartz vein by a company with Mr. Ben Hall, recently of the Hydrographic Division of the United States Geological Survey, in charge. Several shafts were sunk and some drifting done in the course of this work. A stamp mill was erected on the property and it is reported that about eight thousand dollars worth of gold was secured. In 1905 a shaft was sunk by another management close to the old works to a depth of a hundred feet and the vein intersected at about ten feet from this shaft by a cross cut. According to reports, the vein, as

FIG. 1.—VIEW OF A PORTION OF THE WEST OR UPPER CUT, FINDLEY GOLD MINE,
DAHLONEGA, GEORGIA

FIG. 2.—AURIFEROUS QUARTZ STRINGERS IN PARTIALLY DECOMPOSED ROCK, EAST OR LOWER
CUT, FINDLEY GOLD MINE, DAHLONEGA, GEORGIA

exposed in a thirty-foot drift is rather small, but at this level carries high values. Nearer the surface the vein is said to have an approximate thickness of three feet. As none of the old underground works were accessible at time of visit the ore could not be inspected nor any samples secured for assay.

Something like two-thirds of a mile northeast of the Warren mine, a shaft, known as the Allen shaft, was sunk about 1889 by Mr. Sim Lane and others to an approximate depth of fifty feet on a quartz vein several feet in thickness. Two or three additional shafts have been sunk on different veins on the property connected with the Warren mine, but nothing very definite can be stated concerning the character of the ore bodies. Some prospect work has also been done on a body of land a short distance northwest of the Warren mine known as the Wilson and Watson tract. The Warren mine is owned by Mrs. A. L. Sellers, of Baltimore, Md.

McDUFFIE COUNTY

The McDuffie belt takes its name from this county and attains here its greatest importance. It traverses the extreme northwest portion of the county with a southwest-northeast course and has an average width of not over two miles. The town of Thomson, on the Georgia Railroad, is the most conveniently situated station to this portion of the belt.

Deep mining has been conducted successfully in McDuffie county for many years. It is claimed for this section that the earliest discovery of gold in Georgia was made here, and that one of the first stamp mills operated in the United States was erected on Little River in the vicinity of the Columbia mine.[1] Few, if any, counties in the State offer a more attractive field for permanent operations in underground mining than McDuffie. The auriferous quartz veins, while exhibiting pinches and

---

1. Fluker, W. H., Trans. Amer. Inst. Min'g. Engineers, 1903, Vol. XXXIII, pp. 119-125.

swells, are characterized usually, by distinct walls and appear to have been formed in rather clean cut fissures. At some of the mines a "gouge" of an inch or so in thickness occurs between the veins and the walling. Chalcopyrite and galena are frequently present in the vein-quartz along with the more predominant iron pyrite, and pyromorphite has been noted at one or two localities in the oxidized portions of the veins.

The gold belt throughout its entire course in this county has a varying approximate distance from the Coastal Plain of from twelve to fifteen miles. The topography of the Piedmont Plateau at this locality exhibits slight relief as compared with that of some of the gold-bearing regions of the more northern portions of the State. Rock exposures, therefore, are rather limited and opportunities for studying structural relations are not as numerous here as at higher altitudes.

The few specimens of fresh rocks obtained from mine workings in this region and described in the text show fine grained schists and gneisses most probably of igneous origin. An area of granite occurs in the western part of the county and granite-gneiss has been noted within four or five miles of the gold belt. The most characteristic rock of the belt is a dark gray, very fine grained sericitic mica schist. Good examples are to be seen at the Columbia and Parks mines. All of the rocks studied from the McDuffie belt show evidence of having been subjected to great shearing and crushing, and they have, in most cases, been profoundly metamorphosed. At the Parks mine the cleavage surfaces of the schist frequently exhibit corrugations.

EDWARDS, BALBACH AND GERALD MINES.—These properties are situated in the southwest part of the gold belt in McDuffie county near the Warren county line. A limited amount of vein mining, principally in the nature of prospect work, has been done at all three of the localities. No work was in progress at the time of visit and nothing definite can be stated concerning the character of the deposits. The Edwards mine is controlled by Mr. T. A. Scott, of Thomson, Ga., the Balbach, by Mr. E. J. Balbach, of Newark, N. J., and the Gerald, by Messrs. Ira Farmer and W. D. Story, of Thomson, Ga.

HAMILTON MINE.—This mine is situated about a half a mile northeast of the Thomson public road and a few hundred yards southwest of the Columbia mine, described in succeeding pages. The more important mining operations that have been carried on here were conducted in 1901 and 1902 by the Hamilton Mining Company. A working shaft, situated a few yards west of the boundary line of the Columbia mine tract, was sunk on a vein thought to be connected with the principal vein of that mine. No work has been in progress for a number of years and as the underground works could not be entered at the time of visit nothing definite could be ascertained concerning the character of the deposit. It is reported that considerable drifting was done at eighty, one hundred and twenty, one hundred and sixty and two hundred-foot levels. Several auriferous veins are stated to have been cut in the course of these mining operations. In addition to the work above described, a shaft was sunk on a vein on this property known as the Bell vein.

The Hamilton mine has the reputation of having yielded ore of a good grade and, being situated close to auriferous deposits that have been successfully and extensively mined, it was regretted that opportunity was not afforded to make a detailed inspection of the underground works. The property is controlled by Mr. George F. Chamberlain, of New York.

COLUMBIA MINE.—The Columbia mine is situated on a tract of land known as the Forty-acre Lot, lying immediately northeast of the Hamilton mine. The locality is about eleven miles northwest of Thomson, the county seat, on the Georgia railroad.

This mine has been one of the most extensively operated deep gold mines in the State. For a number of years prior to the Civil war, underground work of a rather extended character, for that period, was conducted on the more important ore bodies. After being operated in later years at intervals by different parties, it was purchased in 1899 by the present Columbia Mining Company, under whose management the most important developments have taken place. When the property was visited in the spring of 1908, in the course of field work preparatory to this report, the mine was closed. From inspection of the larger portion of the underground works by the writer, several years previous to that time, and from data obtained from a special report on this property made by the late State Geologist, Prof. W. S. Yeates, in 1902, the following concerning the deposits is given:

Several auriferous quartz veins have been mined on the Forty-acre Lot. The most extensively exploited one is known as the Columbia vein. This vein, where it has been mined, occurs in a fine grained sericitic mica schist, the rock previously spoken of as being the most characteristic rock of the McDuffie belt. A few pieces of a coarse grained more basic rock, found on microscopic examination to contain considerable amounts of hornblende, were also noticed on the dump. As the latter was not observed in place its relations to the mica schist can not be stated. The vein consists of solid milky white quartz and the contained sulphides, pyrite and galena with smaller amounts of chalcopyrite. Pyromorphite is found occasionally in the oxidized portion of the vein. Free gold in the quartz dissociated from the sulphides is not uncommon. The general strike of the Columbia vein is northwest and it dips to the northeast at an angle of about 45°. Its thickness shows variations at different points from less than a foot to several yards. About two hundred feet northwest of the main shaft a vein of considerable size branches off from the Columbia vein toward the north. This vein is designated as the North vein. The main working shaft of the mine is a five by ten foot, two-compartment shaft, sunk on the incline of the vein. At the time of visit, this shaft had an approximate depth of four hundred and fifty feet on about a 45° incline. From the main shaft drifts have been driven in either direction along the vein at different levels. Old workings on the ore bodies made many years ago may be considered the first, or uppermost, level. These have not been reopened by the present management. The second level has a length of six hundred and seventy-five feet to the northwest and a hundred and fifty to the southeast. The third level has been driven four hundred and ten feet to the northwest and a hundred and sixty-two feet to the southeast. The fourth level has been run northwestwardly only and has a length of four hundred and fifty feet.

In describing the Hamilton mine, it was stated that the principal vein there was supposed to be connected with the Columbia vein. The Columbia Mining Company has prosecuted some work on this vein west of the Columbia mine on their side of the line between the two properties. Three drifts have been driven to the southeast; one at a hundred and twenty-foot level, for a distance of a hundred feet; one at a hundred and sixty-foot level, for a distance of a hundred and sixty feet; and one at a two hundred-foot level, for a distance of ninety-five feet.

Approximately three hundred yards northwest of the Columbia mine, considerable work was done some years ago on a vein known as the Bell vein. The principal shaft is a hundred and forty feet deep and several hundred feet of drifting has been done at two levels. This shaft wãs not accessible at the time of visit, but the vein, as seen for a short distance in one of the drifts when the property was visited a few years ago, showed a thickness of several feet. Mr. W. H. Fluker, general manager of the Columbia Mining Company, estimates the thickness of this vein, where it has been exploited, as varying from two to eight feet. The general trend of the Bell vein is northeast and southwest, while the Columbia vein, as previously stated, has a general northwestward strike. It is not improbable, therefore, that the two veins may eventually be found united at some point to the west. In addition to the characteristic dark gray mica schist, a light colored gneiss occurs at the Bell shaft. According to Mr. Fluker, the vein is along the contact of the two rocks.

Microscopic study and chemical analysis of the gneissoid rock show it to have been derived from a granite, or granite porphyry, more basic in composition than the average of the Georgia granites. Feldspar, occurring principally as phenocrysts, is in excess of quartz and small flakes and shreds of biotite, largely altered to weakly pleochroic chlorite, are scattered throughout the slide. Reference to the chemical analyses of this rock and that of the dark gray sericitic mica schist, taken from the lowest level of the Columbia mine, which are given in Chapter IV, will show the close identity in the percentage composition of the two. While appearing quite different, both in the hand specimen and under the microscope, this similarity in chemical composition suggests that the sericite schist is a highly sheared phase of the gneiss, the prominent feldspars of the latter rock having been crushed up and altered to the large amounts of sericitic mica character-istic of the schist.

In addition to the veins already mentioned, others are said to occur on the Forty-acre Lot. The mining conducted on them, however, has been limited, so far, to prospect work.

No opportunity was afforded at the time of visit to sample any of the veins above described. At the time Prof. W. S. Yeates published the special report previously mentioned, he took several samples from the Columbia vein and one from an ore pile at the mouth of a shaft on the Bell vein. The results of the assays made at the N. P. Pratt Laboratory, Atlanta, and published by him are here quoted.

"Columbia Vein. Sample from ore pile, taken out in the course of sinking the main working shaft, from a hundred and seventy-five to two hundred feet:

Original, 0.28 oz. ($5.60) of gold per ton.

Duplicate, 0.38 oz. ($7.60) of gold per ton.

"Sample from the west breast of the three hundred and ten-foot level (the drift in that direction at the time the sam-ple was taken was about a hundred and ten feet long):

0.77 oz. ($15.40) of gold per ton.

"Sample from the west end at that time of the hundred and forty-foot level eastward for approximately a hundred and fifteen feet. Sample taken both in the drift and stopes:

0.09 oz. ($1.80) of gold per ton."

As the last assay shows a much smaller value than any other assays given by Prof. Yeates, and as Mr. Fluker, the general manager, states that an average of twenty-three head samples, taken when the ore from this part of the mine was milled, was $6.04, it seems probable that this sample failed to represent average values.

Sample from the same drift as the one last quoted com-mencing at a point about eighty feet east of the section men-

tioned in giving that assay and taken in the same manner as
the former one, gave:

0.29 oz. ($5.80) of gold per ton.

A sample taken from the north vein at the hundred and
forty-foot level, at a point ten or fifteen feet from its juncture
with the Columbia vein, gave:

0.52 oz. ($10.40) of gold per ton.

The vein, where sampled, showed a thickness of about four
feet.

In the hundred and sixty-foot level on the Hamilton vein,
Prof. Yeates took a sample for assay from a vein that he considered the Columbia vein. The sample was taken at a point
where the vein was exposed in the drift for about ten feet on
its hanging wall side and for about four feet in a breast. The
results of the assay of this sample show:

5.00 oz. ($100.00) of gold per ton.

One assay is given from a sample of the Bell vein ore taken
from an ore pile at the mouth of a shaft on that vein designated as shaft No. 7. This assay shows:

0.43 oz. ($8.60) of gold per ton.

In the same report from which the foregoing assay results
were taken, United States mint returns of gold shipped from
the Columbia mine during the year 1899, 1900 and 1901 are
given. The returns for each of these years, after being consolidated into totals from the different shipments as there
stated, are given below:

For the year 1899_____$4,837.90
For the year 1900_____ 7,322.88
For the year 1901_____ 7,272.76

The milling plant of the Columbia mine is situated close to
the principal working shaft. (Plate VII, Fig. 2.) The
tramway of the shaft is extended on about the same incline

from the collar of the shaft to the upper story of the mill. The ore is received at the mouth of each level and elevated by a skip, without subsequent unloading, to the mill. The mill consists of a ten-stamp battery of seven hundred and fifty pound stamps. The free gold is secured by amalgamation and the sulphides saved with two Wilfley concentrating tables. The concentrates are not treated at the mine. Air drills are used in the underground work and the mine is supplied with an electric plant for lighting purposes.

MOTES TRACT.—About a half of a mile northwest of the Columbia mine some mining operations have been conducted on auriferous quartz veins. This work, done several years ago, consisted principally of prospect work and nothing very definite can be stated concerning the character of the deposits.

HENRICH PROPERTY.—This property lies immediately east of the Motes tract. At a point about three-fourths of a mile north of the Columbia mine some prospect work has been done on a quartz vein by Mr. Carl Henrich, of New York City. This work was of a very limited character and nothing definite was ascertained concerning the value of the deposit.

PARKS MINE.—The Parks mine is something less than a mile northeast of the Columbia mine. The locality, as is the case at the Columbia mine and other adjoining properties, is in a section having a rolling topography and well supplied with water and timber. External conditions for mining are excellent.

Mining for gold was done here prior to the Civil war, but no records are obtainable as to the results of these early workings. Later, Mr. J. Belknap Smith conducted mining operations at this mine until the time of his death in 1888. For about ten years thereafter, his widow, Mrs. J. Sep Smith, carried on mining work on a quartz vein at this locality. The total amount of work done during that period was quite limited, but the results were very satisfactory and added greatly to the reputation of the mine as a gold producer. The principal mining operations carried on by Mrs. Smith were confined to a quartz vein occurring a few rods to the southwest of one on which work was being conducted at the time of visit. The main shaft on Mrs. Smith's vein was known as the Water Shaft and was sunk to a depth of a hundred and fifty feet. Several levels were driven from this shaft, the longest being at the bottom and having an approximate length of two hundred and fifty feet along the south extension of the vein. These underground works have been inaccessible for a number of years, but the vein is reported by those familiar with the mine to consist of solid quartz heavily impregnated with pyrite, galena and chalcopyrite. It is stated that the dip is nearly vertical and that the strike is approximately north and south, cutting the trend of the country rock. The vein where exploited is said to have an average thickness of several feet.

The ore taken out was milled at a stamp mill located about three miles distant on Little River. Only the gold that could be secured by amalgamation on the plates was saved, the sulphides being allowed to go to waste in the river.' The United States mint returns of the gold obtained during the course of these mining operations have been preserved by Mrs. Smith and were kindly submitted for inspection. Considering the limited amount of mining done and the neglect to save the gold in the sulphides they indicate that the values in the vein must have been considerably higher than the average of the gold ore that has been mined in Georgia. High-water mark seems to have been reached in 1892, for which year the returns show $11,363.23 after deducting coinage charges. Of this amount, $50.49 was for silver values.

A few rods to the northeast of the vein exploited by Mrs. Smith, as above stated, another auriferous quartz vein, known as the Parks vein, occurs on which mining operations have been conducted at different times. In addition to earlier operations, this vein was worked under a lease by the Georgia-Montana Gold Mining Company for a limited period a few years ago. Under this management the present mining plant of the Parks mine was built and equipped.

At time of visit, work had been renewed on the Parks vein by Mr. W. H. Fluker of the Columbia mine whose tireless energy and intelligent efforts have been devoted for a number of years to building up the gold industry in McDuffie county. A working shaft was being sunk and the old drifts cleaned out, but these operations had not progressed far enough to allow the vein to be inspected and sampled. During a visit to the property several years ago, when it was being operated by the company previously mentioned, a portion of the underground works was traversed and the vein examined for a hundred feet or more in one of the drifts. As there seen, the vein was a very clearly defined one, showing a thickness of several yards for the greater portion of the exposure and consisting of solid quartz carrying pyrite, galena and chalcopyrite in amounts that varied at different points.

Since the close of the past year, 1908, Mr. Fluker, general manager, has furnished a copy of stamp mill reports for that year from which the following data are given. In the figures following, a small percentage of silver values is included with the gold.

Ore Milled at Parks Gold Mine, During the Year 1908.

March 20	....	48½	tons	$ 192.30	......$ 3.96	per ton.
May 18	......	42	``	304.44	...... 7.24	`` ``
June 12	....	51	``	283.39	...... 5.55	`` ``
July 7	....	146	``	865.58	...... 5.92	`` ``
July 15	......	45	``	324.73	...... 7.21	`` ``
July 24	......	100	``	626.50	...... 6.26	`` ``
Aug. 10	......	153	``	854.24	...... 5.58	`` ``
Aug. 20	......	125	``	1,729.65	...... 13.83	`` ``
Sept. 5	....	51	``	529.44	...... 10.38	`` ``
Sept. 18	......	96	``	696.27	...... 7.25	`` ``
Sept. 29	......	48	``	394.00	...... 8.20	`` ``
Oct. 9	....	100	``	684.40	...... 6.84	`` ``
Oct. 21	....	50	``	424.14	...... 8.48	`` ``
Nov. 3	....	110	``	928.56	...... 8.44	`` ``
Nov. 29	......	95	``	410.00	...... 4.32	`` ``
Dec. 15	......	68	``	497.22	...... 7.31	`` ``

The total hours of mill run were 1,895. From these figures it will be seen that 1,328 tons of ore were milled yielding $9,744.86 in free gold, an average of $7.33 of free gold per ton. United States mint returns were exhibited verifying the values above given, excepting the figures under date of November 29th. Mr. Fluker estimates the value of the concentrates saved at $2,067.00. Combining the values saved of free gold and the estimated value of the concentrates and the tailings, Mr. Fluker places the total average value of the ore at $14.71 per ton. His report gives the average thickness of the vein, where mined. as fourteen feet.

It will be seen from the figures above given, that the mill was operated, in round numbers, seventy-nine days of twenty-four hours each during the year and that the average number of tons of ore milled per day was a little less than seventeen, the average yield per day of free gold being approximately $123.30. The results of these limited operations, considered in connection with the previously described work successfully prosecuted a number of years ago on the contiguous Smith vein, afford encouraging prospects for more extensive mining in the future at this locality.

In addition to the veins already mentioned, it is reported that several others occur on the property of the Parks mine and that a limited amount of work has been done on them in the past, but nothing definite can be stated concerning their character and value.

The milling plant of the Parks mine, situated contiguous to the Smith and Parks veins. consists of a ten-stamp mill, two Johnson concentrators, a forty-horse-power engine, boilers, etc.   (Plate VI. Figs. 1 and 2.)

LANDERS PROSPECT.—About a fourth of a mile north of the Parks mine, two prospect shafts a few yards apart were sunk in 1907 on an auriferous quartz vein, or veins.   No outcroppings were noticed at the surface and the deposits could not be inspected in the shafts at the time of visit.   An ore pile of about five tons at the locality, showed specimens of vein quartz characteristic of the auriferous veins of the region.   Pyrite and galena were present in considerable quantities, together with chalcopyrite in less amounts.   Free gold was noticed in several pieces of the ore.   An assay sample taken from this ore pile showed a value of $15.71 per ton.

The property is controlled jointly by Mr. W. H. Fluker, of the Columbia mine, and Mr. W. D. Story, of Thomson, Ga.

PORTER MINE.—The Porter mine is situated about three-fourths of a mile north of the Parks mine.   Several auriferous quartz veins occur at this locality.   Two, known respectively as the Brock and the Egypt veins, have attracted the most attention and mining operations have been confined principally to them.   A very limited amount of work has also been done on another vein designated as the Island vein.   No work has been in progress at the Porter mine for several years and nothing very satisfactory can be stated concerning the character of the deposits.

Mining operations were conducted on the Brock vein several years ago for a short period by the National Mining Company. The main shaft has caved at the surface. The vein is said to strike nearly east and west.

The principal shaft on the Egypt vein is stated to be about a hundred and thirty-five feet deep and several hundred feet of drifting has been done. A mill, equipped with ten stamps, is located at the Egypt shaft. The property is controlled by Mr. L. J. Peck, of Boston, Mass.

TATHAM MINE.—The Tatham mine is situated approximately a mile and a half northeast from the Parks mine and close to a road that branches to the north from the Washington-Augusta public road near Raysville.

Considerable underground mining has been conducted here in the past, but no work of any importance has been in progress for about fifteen years. The principal shaft, known as the Tatham shaft, has caved at the surface and no exposures of the vein were noticed. It is reported that some very rich ore was secured when the mine was in operation. A mill, equipped with ten stamps and a Wilfley concentrating table, is located near the shaft. The property is owned by the Tatham estate of Philadelphia, Penn.

WOODALL MINE.—This mine is about a half a mile northeast of the Tatham mine and close to Flint Hill church on a branch of the Washington and Augusta public road. No mining operations have been carried on at the locality for some years and no exposures of the vein were noticed. The working shaft is a two-compartment one, said to have a depth of about a hundred and ten feet. It is reported that the vein where exploited had an average width of something like eighteen inches and afforded ore of a good grade. The property is controlled by Mr. M. L. Bickart, of Baltimore, Md., and Mr. I. H. Oppenheim, of Atlanta, Ga.

GRIFFIN MINE.—The Griffin mine is situated about a mile and a half northeast of the Woodall mine and close to Little River. Considerable mining has been carried on at the locality in the past, but no operations of any magnitude have been in progress for a number of years. Nothing very definite can be stated concerning the character of the deposits.

RAYSVILLE BRIDGE VEIN.—One or two hundred yards above the public bridge across Little River at Raysville a quartz vein of considerable magnitude outcrops strongly in the bed of the river forming a slight shoal at that point. With the assistance of Mr. M. A. Pearce, of Thomson, whose property borders one side of the stream at the locality, a sample for assay was taken from this vein where exposed in the river. The sample on being assayed yielded no gold. Owing to high water, however, at time of visit, only a very limited exposure of the vein was accessible and it was regretted that a more satisfactory test could not be made.

WILKES COUNTY

A very small area in the southeast corner of this county is traversed by the McDuffie gold belt. Mining operations on the Wilkes county portion of the belt have been limited in extent.

EDMUNDS MINE.—This mine is about a mile and a half from Amity, in Lincoln county, and immediately adjacent to the Wilkes-Lincoln county line. The major portion of the tract of land the mine is situated on lies in Lincoln county. The Edmunds mine has been operated on a small scale by different parties at different times since early in the eighties.

No mining operations have been conducted at the locality in recent years. Several shafts have been sunk on an auriferous quartz vein occurring here, and open cut work and drifting done. It is reported that the vein has been mined to water level for an approximate distance of four hundred feet along its strike. No exposures were noticed, but judging from the line of works the strike of the vein is nearly east and west and it is said to dip to the north. The main shaft is about ninety feet deep and drifts have been driven from the bottom for something over a hundred feet in both east and west directions.

In 1896-97-98 and 99, Messrs. S. R. and J. F. Edmunds, of Amity, Ga., the owners of the property, conducted mining operations of a very limited character on the vein. A portion of the United States mint returns for the gold shipped during that period has been preserved and was submitted for inspection. Those of the returns preserved showed a total of $1,175.99.

About a hundred yards to the north of the vein on which the principal mining operations have been prosecuted, another vein occurs, which has been prospected to a limited extent.

HILLY MINE.—This mine is situated in the southeast corner of the county about a mile and a half south of the Washington-Augusta public road. A little mining was done here some years ago on what appears, judging from very limited exposures, to be an auriferous zone consisting of a schist impregnated with pyrite and containing small quartz veins or stringers of quartz. A shaft was sunk during the course of these operations to a depth of eighty feet, and a little drifting done. The ore is reported as being of low grade. The mine is controlled by Mr. R. K. Reaves, of Athens, Ga.

LINCOLN COUNTY

The McDuffie belt traverses the southern portion of Lincoln county from Raysville, at the McDuffie, Wilkes and Lincoln county corners, to the Savannah River at a point a little south of Double Branches and about three miles above the mouth of Little River.

Gold mining has been carried on in this county to a much more limited extent than in McDuffie. Ore of good grade, however, has been milled at two or three localities. The general character of the ore furnished by the auriferous quartz veins is similar to that of the ore obtained in the last mentioned county. Pyrite and galena are the most commonly occurring sulphides in the veins and chalcopyrite is frequently encountered.

RIVERS PROPERTY.—Some prospect work has been done on a quartz vein on this property at a point about a fourth of a mile east of Raysville. A shaft was sunk on the vein to a depth of seventy feet at the line between this property and the Ramsey property on the east. In an old shaft a little southwest of the one just mentioned the vein showed a width of about a foot. An assay sample taken from the vein as exposed in the shaft and assayed at the N. P. Pratt Laboratory, Atlanta, showed a value of $0.80 per ton. The general strike of the vein is N. 40° E.

RAMSEY PROPERTY.—This property lies immediately to the east of the Rivers property. Several auriferous quartz veins have been located on the tract and some prospect work carried on at different times. The vein described as occurring on the Rivers property has had some work done on it close to the line between the two tracks. Between an eighth and a quarter of a mile further to the northeast an inclined shaft was sunk on a vein that may be a continuation of the Rivers vein. A little drifting was done from this shaft and a limited amount of ore taken out and milled at a small stamp mill that was erected on the property. No record was kept of the gold obtained. The vein where exposed in this shaft showed a width of from two to three feet, but a considerable amount of wall rock was interlaminated with the quartz. Several other prospect shafts have been sunk on the vein in addition to the one just mentioned.

About fifty yards to the north of the above described vein another quartz vein occurs on which a shaft was sunk several years ago, to a depth of twenty feet. The vein at the bottom of this shaft showed a thickness of about two feet, and consisted of solid quartz carrying pyrite and small amounts of galena and chalcopyrite. Free gold was noticed in some specimens of the ore. An assay sample taken from the vein at this shaft and assayed at the N. P. Pratt laboratory, Atlanta, yielded $4.80 of gold per ton.

PASCHAL MINE.—This mine is situated on Hickory-nut Hill, a short distance to the east of the Ramsey property. An auriferous quartz vein was discovered at this locality and a little prospect work done before the Civil war. Regular mining operations, though limited in extent, were conducted on the vein during the year 1901, by the Clay Hill Mining and Milling Company. The ore taken out was milled at a small stamp mill erected on Loyd's Creek about a mile from the mine. At the time of a visit to the mine in the fall of 1901, the underground works consisted of a shaft a little to the southeast of the crest of the hill, having a depth of about twenty-four feet, on the vein, and connected by a drift about thirty-five feet in length along the northeast extension of the vein with another shaft, known as the Bell shaft, having a depth of forty feet. A drift about fourteen feet long had also been run southwest along the extension of the vein from the bottom of the twenty-four foot shaft. At the surface the vein showed a thickness of something less than a foot, but at the bottom of the Bell shaft the thickness was approximately four feet.

Near the surface the vein exhibited sulphides in sparing amounts only, but at the bottom of the Bell shaft the characteristic sulphides of the McDuffie belt were present in considerable quantities, galena being especially abundant. Two samples for assay were taken at this time: No. 1, from the two drifts above mentioned, and No. 2, from the bottom of the Bell shaft. These samples, assayed at the N. P. Pratt laboratory, Atlanta, yielded the following results:

> No. 1_____0.41 oz. ($ 8.20) of gold per ton.
> No. 2_____1.69 oz. ($33.80) of gold per ton.

In 1902, the mine was operated for about a year by Messrs. D. C. Stainback and T. B. Wiley. A shaft was sunk to a depth of eighty-five feet about thirty feet northwest of the Bell shaft and connected with the latter by a winze. A stamp mill and engine were installed on Hickory-nut Hill at the mine. During a visit to the locality at this time it was noticed that an enlargement occurred in the vein at the bottom of the new shaft similar to the enlargement spoken of as occurring at the bottom of the Bell shaft. Later, the mine was operated for a short time by Mr. F. H. Hyatt, of Columbia, S. C.

The assay tests made, together with satisfactory information obtained as to the amount of gold secured from very limited operations, indicated the occurrence at this mine of some ore of good grade. It was impossible, however, when the underground works were last examined to arrive at any very satisfactory conclusions as to the probable average thickness of the vein for a considerable horizontal or vertical extent.

DILL PROPERTY.—This property adjoins the Paschal mine on the east. At the time of visit some prospect shafts had been sunk on quartz veins on this property, but the work had not progressed far enough to admit of arriving at any conclusions as to the probable value of the deposits.

JULIA, OR PHELPS, MINE.—This mine is located near the S. K. Dill property and about a half a mile north of Little River. The property was prospected several years ago by James Frank and Son, of Augusta, Ga., and three or four auriferous quartz veins were located and exposed by them in pits or shafts. The deepest openings were made on a vein in the northeast part of the property designated as vein No. 1. The two other veins near each other were found in the southern part of the property and designated as veins Nos. 2 and 3. The property was visited in company with Mr. Edward Frank soon after this prospecting was done and the veins examined and samples taken for assay. Since that time the mine has passed into the possession of Mr. Walter Phelps, of Baltimore, Md., who erected a small stamp mill on the property and prosecuted additional development work. When the McDuffie gold belt was traversed during the course of field work in 1908, no mining operations were in progress at this locality and it was not revisited. At the time of examination previously mentioned, vein No. 1 had been exposed by Mr. Frank in two shafts and several shallow cross cuts. In the south shaft, which was about fifteen feet deep, the vein showed a thickness of approximately a foot near the surface and two feet at the bottom of the shaft. The north shaft was twenty-five feet deep when examined. Near the surface the vein was about ten inches in thickness, and widened gradually to two feet or more at the bottom of the shaft. The general strike of the vein as indicated by the line of openings is about N. 10° E, with an easterly dip. The vein quartz contained considerable amounts of iron pyrite and galena. The two veins spoken of as occurring in the southern part of the property had been exposed at time of visit in several shallow pits or shafts. Vein No. 2 was open to inspection in only one shaft, the others being full of water. As seen in this shaft, which was fifteen feet deep, the vein had a maximum thickness of about fifteen inches at the bottom of the shaft and was thinner at the surface. Vein No. 3 showed a width in several pits towards the south of about seven inches and the quartz contained large amounts of iron pyrite and considerable chalcopyrite. In one pit towards the north, the vein was about two feet wide, but badly shattered. In the more northerly situated pits the vein quartz contained less sulphides than that examined in the pits to the south.

Two samples for assay were taken from vein No. 1 and one sample from vein No. 2 and one from vein No. 3. Sample No. 1 from vein No. 1 was taken from the north shaft, and sample No. 2, from the south shaft. The sample from vein No. 2 was taken from the south shaft. The sample from vein No. 3 was taken from several pits along the strike of the vein. These samples, assayed at the N. P. Pratt laboratory, Atlanta, yielded the following results:

> Sample No. 1, vein No. 1_____$18.80 per ton.
> Sample No. 2, vein No. 1_____$ 1.40 per ton.
> Sample from vein No. 2_____$ 1.20 per ton.
> Sample from vein No. 3_____$22.00 per ton.

Mr. Frank stated that several other veins in addition to those described had been located at this locality, but, owing to heavy rains at the time of visit, the prospect pits were flooded and they could not be inspected. As a number of veins occur contiguous to each other on this property and as the average value of the very limited assay tests above given is $10.85 per ton, the prospects for successful mining here would seem to be rather encouraging.

S. K. DILL PROPERTY.—This property lies close to the Julia mine tract. Several quartz veins were noticed outcropping in the vicinity of the Dill residence, but up to the time of visit no mining developments had been undertaken.

OTHER PROPERTIES IN LINCOLN COUNTY.—Between the last described properties and the Savannah River, the occurrence of gold has been noted at several localities along the course of the belt, but at the time of visit few mining developments meriting description were observed.

About a fourth of a mile from the Savannah River, and approximately a half of a mile from the Petersburg public road, some old works were noticed on the Bussey property. Several shafts, now nearly filled up, are to be found, and it is reported that considerable work for gold was done at the locality many years ago. Nothing could be learned as to what success attended these mining operations.

THE OGLETHORPE BELT

OGLETHORPE COUNTY

The Oglethorpe gold belt traverses the southeastern part of the county. Commencing at a point about three miles northeast of Bairdstown in the southwest part of the county, the course of the belt is northeast to the vicinity of Goose pond near the Elbert county line. Mining operations have been confined, so far, to the southwest half of the belt as here defined, very little being known as to the extent and probable value of the deposits in the northeastern portion.

An interesting mineralogical occurrence at several mines in this gold belt is the presence of native sulphur in the oxidized portion of the auriferous quartz veins. The sulphur is found in a powdery form in cavities in the quartz originally occupied by iron pyrite.

BRISCOE MINE.—This mine is situated about three and a half miles northeast of Bairdstown, a station on the Athens branch of the Georgia Railroad. A limited amount of mining consisting principally of open-cut work, was done here prior to the Civil war. A considerable excavation was made in the course of these operations on an auriferous quartz vein, but no exposures were found in the old cut and the ore could not be examined.

DRAKE MINE.—The Drake mine is situated about a fourth of a mile northeast of the Briscoe mine. A limited amount of mining has been carried on at this locality at different times and by different parties. Several prospect shafts have been sunk on an auriferous quartz vein and some years ago ore was hauled to a small stamp mill situated on a creek in the vicinity of the mine, and some milling done. It is reported that very **good** returns were obtained from the ore milled.

At the time of visit some prospect work was in progress at the locality under the direction of Mr. J. C. Foreman, of Savannah. A shaft was being sunk on the vein a short distance from some old workings. This shaft, which was about sixty feet deep when examined, did not afford a very good opportunity for examining the vein, but about fifty feet from the surface the latter had been cut into from one side for something like three yards without the entire thickness being exposed. The vein, where seen, consisted of solid quartz with rather sparing amounts of sulphides. A sample for assay was taken from the exposure in the shaft above mentioned and also from an ore pile of a number of tons at the mouth of the shaft. These samples on assay yielded the following results:

> Sample from shaft _____$0.41 per ton
> Sample from ore pile _____ 2.06 per ton

BUFFALO, OR HOWARD, MINE.—The Buffalo mine is situated about five miles northeast of the Drake mine and six and a half miles east of Stephens, a station on the Athens Division of the Georgia Railroad. Some mining on a limited scale was carried on here about 1899 and 1900 by Messrs. Charles Howard and Sim Lane, of Oglethorpe county. The locality is on a hillside near Buffalo Creek. A small stamp mill was erected and it is estimated that several thousand dollars were secured in the course of mining operations. Several auriferous quartz veins have been prospected on the property and some specimens of ore quite rich in free gold have been exhibited from this mine. At time of visit the old shafts and pits had caved to such an extent that very little could be ascertained in regard to the character and extent of the deposits. When the property was examined, Mr. Charles Howard, of Stephens, Ga., was the principal owner.

FLUKER-STORY PROSPECT.—Approximately a mile southeast of the Buffalo mine and immediately adjoining the Guarantee mine, described below, Messrs. W. H. Fluker and W. D. Story did some prospect work about ten years ago on the Johnson property. A quartz vein was stripped at the surface for something like a hundred yards along its strike, the general course of which is N. 10° E. The vein, as exposed, showed an average thickness of about a yard and carried considerable amounts of sulphides. The work had not been prosecuted to an extent sufficient to permit any very satisfactory conclusions to be drawn as to the probable value of the deposit.

GUARANTEE MINE.—This mine is situated immediately northeast of the Fluker-Story prospect. Some open cut work was done on a quartz vein on this property before the Civil war by Mr. John Wynn, who appears to have been the pioneer of the gold mining industry in Oglethorpe county. About 1900, the Guarantee Mining Company sunk a shaft to an approximate depth of a hundred and twenty feet and did some drifting on a quartz vein that is probably a continuation of the vein on the Johnson property above mentioned. This shaft could not be entered at time of visit. Some pieces of vein quartz lying about its mouth showed considerable amounts of iron pyrite. A tunnel was also driven into the hillside from a point near a small branch. The vein or a branch from it was exposed for a short distance in this tunnel. The company also erected a ten-stamp mill near the mine, but no record was available as to the amount of gold that they secured. A year or so later a little additional work was done at the locality by Mr. E. P. Cowell, of Philadelphia, Penn. The mine was owned at the time of visit by Mrs. S. S. Cowell, of Philadelphia, Penn.

MORGAN MINE.—The Morgan mine is situated about two miles northeast of the Guarantee mine and six and a half miles southeast of Lexington, the county seat of Oglethorpe county. Much more extensive gold mining operations have been carried on here than at any other locality in the county. Mr. John Wynn, the original owner of the property, erected a stamp mill and prosecuted work at this mine on an auriferous quartz vein before the Civil war. Later, the mine was worked at intervals by several different managements. including the Morgan Mining Company, the Lexington Gold Mining Company, and others. At time of visit, the property was under the control of Mr. W. E. Shem, of Alliance, Ohio.

The principal mining operations at the Morgan mine have been conducted on an auriferous quartz vein having a general strike of N. 12° E. and a nearly vertical dip. The two more important shafts are about four hundred feet apart and are designated respectively as the South, or Water shaft, and the North shaft. Mining operations in recent years have been carried on entirely from the South shaft. Approximately a hundred yards of open cut work has been done on the vein between the two shafts. At time of visit the South shaft had a depth of something over a hundred feet. An old drift at a sixty-foot level extends from this shaft northeastward along the strike of the vein for about a hundred and seventy-five feet and has a southwestward extension of about forty feet. A drift has also been run for a considerable distance northeastward from the bottom of the shaft. As exposed in the South shaft the vein will average something like four feet in thickness and in the upper portion of the shaft is composed principally of solid quartz, but in the lower portion contains considerable interlaminated wall rock. As seen in the level at the bottom of the shaft, the vein exhibited a number of breaks and irregularities in thickness and in this particular drift as far as the vein had been exposed the prospects of a probable extensive supply of ore were not very encouraging.

At the time of visit, owing to some defect in the pumping machinery, opportunity was not afforded for taking assay samples. During a previous visit made some years ago, when the mine was being operated under a different management, two samples were taken for assay. Sample No. 1 was taken from the southwest extension of the old sixty-foot level and from a section of about sixty feet from the shaft in the northeastward extension of the same drift. Sample No. 2 was taken from the vein as exposed in the bottom of the shaft and also on the sides between the bottom and the sixty-foot level. These samples on being assayed at the N. P. Pratt laboratory, Atlanta, yielded the following results:

No. 1_____.17 oz. ($3.40) of gold per ton
No. 2_____.27 oz. ($5.40) of gold per ton

The chief sulphides of the vein are pyrite and chalcopyrite, and a little below water level, native copper in small masses is occasionally found in seams and cracks in the vein quartz. The wall rock at this mine is a highly sheared and altered igneous rock which in thin section shows a structure and mineral composition suggestive of a grano-diorite. Pheno-crysts both of quartz and feldspar badly crushed up, and in the case of the latter mineral, largely altered to sericite or kaolin, are present in a fine grained ground-mass composed of quartz and feldspar and sericite and considerable amounts of chlorite probably derived from biotite. Much of the quartz and feldspar of the ground-mass, and to a less extent in the case of that of the phenocrysts, exhibit a granophyre structure. Considerable amounts of secondary calcite are present. The analysis of this rock shows a chemical compo-sition corresponding more nearly to a grano-diorite than a true granite.

As previously stated, no mining operations had been in progress, up to the time of visit, at the North shaft for a number of years. It is stated to have an approximate depth of eighty-six feet with a drift at a fifty-foot level that has been run for a distance of sixty feet southwest of the strike of the vein and a hundred feet on the northeast extension of the vein.

The milling plant is situated near the North shaft and is equipped with a ten-stamp battery of eight hundred and forty pound stamps and a Wilfley concentrating table.

OTHER PROPERTIES IN OGLETHORPE COUNTY.—At several localities along the gold belt between the Drake mine on the southwest and the Morgan mine to the northeast, a very limited amount of prospect work has been carried on. None of these operations had progressed far enough at the time of visit to afford any very definite information as to the probable value of the deposits. Work of the above-mentioned character has been prosecuted on the Lumpkin, the Chedel, the Esco, the Brown and the Dunn properties.

WILKES COUNTY

At one point in Wilkes county an auriferous deposit has been mined that may be considered as occurring in the Oglethorpe county belt. The locality is in the northwestern part of the county, close to the Oglethorpe county line and about a mile and a half or two miles southwest of the general trend of the gold belt.

LATIMER MINE.—This mine is situated in the northwestern part of the county on the Danielsville public road. Gold was discovered along a branch close to the Latimer dwelling prior to the Civil war and some sluicing operations of a limited character carried on at the locality. The property came into local prominence, however, in 1901 at which time Mr. W. D. Story of Thomson, Ga., located a small quartz vein, very rich in free gold, on a hill, a short distance from the Latimer dwelling. The vein was only a few inches in thickness and when mined the ore was exhausted at a depth of about thirty-five feet. The ore taken out was milled at the Columbia Mining Company's mill in McDuffie county. According to Mr. W. H. Fluker, general manager of the Columbia mine, about a ton and a quarter of this ore yielded approximately $3,660.00.

Several panning tests of the soil and saprolite material from the surface of the hill in the vicinity of the rich pocket above described indicated the presence of considerable amounts of free gold. The advisability of mining this surface material has been considered by parties interested in the property.

THE MADISON COUNTY BELT

MADISON COUNTY

The Madison county belt traverses a small portion of the southeast part of Madison county. It commences, as at present defined, at a point a little northeast of Comer, a station on the Seaboard Air Line Railway, and runs northeastwardly, entering Elbert county near Carrington's or Moore's ferry on Broad River.

Mining operations have been confined to small placer deposits along branches and creeks, and up to the time of visit little, if any, success had rewarded efforts to locate auriferous veins.

SMITH PROPERTY.—A little work has been done in the past on the C. M. Smith property on the Vineyard Creek public road a few miles northeast of Comer. This property seems to be the most southwestward locality of the Madison belt where any efforts have been made to secure gold.

WEBB, JAMES THRELKELD, MOON AND OLIVER THRELKELD PROPERTIES.—These properties lie a short distance northeast of the Smith property along the course of the gold belt. A little prospect work has been done in gravel deposits along small streams at the several localities. On the Webb tract some mining was done along a branch before the Civil war, the stream having been turned for a short distance from its natural channel.

PARHAM, RIDGEWAY AND KING PROPERTIES.—These properties lie a short distance northeast of the last tracts. On the Parham property gold occurs along a small dry gulch or hollow, near the crest of a ridge about an eighth of a mile east of the Parham dwelling. Several small particles of gold were picked up in the bed of this gulch at time of visit, but as no mining operations had been carried on at the locality no opportunity was afforded to judge the extent of the occurrence.

Gravel deposits along several small streams on the Ridgeway property have been prospected to a limited extent. Some panning tests of gravels at this locality yielded a small amount of gold.

On the King property a little prospect work has also been done. The character of the deposits are similar to those of neighboring properties.

CARRINGTON PROPERTY.—This property adjoins the King property and is about five miles southwest of Bowman in Elbert county. Auriferous gravel deposits occur along several small streams on this property. The deposits along two of these, namely, No Land and Little California creeks, have attracted the most attention. Regular mining operations on a small scale have been carried on at this locality at different times in the past. Some panning tests made at one or two points were fairly satisfactory, but it is doubtful if the deposits are of sufficient extent to warrant other than limited operations.

Commencing at the mouth of No Land Creek at Broad River, a considerable tract of valley land occurs along the lower portion of this stream. In the early nineties, Mr. W. E. Sedlow of Washington, Ga., installed some pumping machinery at the mouth of the creek for the purpose of pumping water from Broad River with which to carry on hydraulic mining. An engine was put up in the creek lowlands several hundred yards above the mouth of the stream and a limited quantity of auriferous gravels raised with a pump and washed in a sluice box. Work had been suspended before time of visit, but Mr. Sedlow reported encouraging results.

ELBERT COUNTY

The Madison county belt extends from the Madison-Elbert county line at Moore's, or Carrington's ferry, on Broad River to a point about three miles northeast of Bowman, a station on the Toccoa-Elberton branch of the Southern Railway.

In character the deposits are identical with those of the Madison county portion of the belt. A very limited amount of placer mining has been carried on in the past along stream courses at four or five localities within a few miles of Bowman.

BROWN PROPERTY.—This property is about three miles southwest of Bowman and immediately adjacent to Madison county. Gold is reported by prospectors as occurring here in gravels along stream courses. For about three miles northeastward from this property along the trend of the belt no auriferous deposits seem to have attracted attention, but it is not probable that a complete gap occurs.

PROPERTIES NEAR BOWMAN.—Within two or three miles of Bowman a limited amount of either placer mining or prospect work has been prosecuted in the past on the James Ginn property, the Marion Hall property, the Hamilton Moss property, the Sanders property and the Pulliam property. The gold obtained at the several localities was from placer deposits along small streams.

THE HALL COUNTY BELT

FULTON COUNTY

The Hall county gold belt, as at present defined, commences in the northern part of Fulton county at a point about seven miles north of the city of Atlanta. A small area in this county is traversed by the belt in a northeast-southwest course. No mining, except a little prospecting, has been carried on in this portion of the belt and the presence of auriferous deposits here is probably more interesting from a scientific standpoint than otherwise.

Gold has been reported as occurring on lots 200, 72 and 38 in the 17th district. At the last mentioned locality some prospect shafts were sunk and a little mining carried on a number of years ago.

MILTON COUNTY

The Hall county gold belt traverses the southeastern part of Milton county. Its position and boundaries have not been as definitely located here as in the counties on its northeast extension. Very little mining for gold has been carried on in Milton county and this portion of the belt has not so far proved of much commercial importance.

A limited amount of mining has been done in the past at a few localities. On Mason's, or Gold lot, in the 1st district, about two and a half miles east of the town of Roswell and three-fourths of a mile north of the Chattahoochee River, placer mining was conducted many years ago along some branches emptying into Seven Branch Creek. In recent years, Mr. George Sciple of Atlanta did a little prospect work for auriferous veins at this locality. About a fourth of a mile from this point on lot 662 near Lizzard Lope school house Mr. Sciple reports the occurrence of considerable placer

FIG. 1.—VIEW OF A PORTION OF THE LARGE CUT OF THE BARLOW GOLD MINE,
DAHLONEGA, GEORGIA

FIG. 2.—VIEW OF A PORTION OF THE MAIN CUT OF THE CROWN MOUNTAIN GOLD MINE,
DAHLONEGA, GEORGIA

gold. Something like a mile northeast of the Gold lot a little placer work was done years ago on a branch emptying into Big Creek. Some prospect work for veins has also been carried on at this point by Mr. Sciple. To the northeast, near Warsaw, some quartz veins were prospected to a limited extent a number of years ago.

FORSYTH COUNTY

The Hall county belt traverses a small area in the southern part of Forsyth county. The portion in which the gold belt is found is a narrow strip of the county only a few miles wide lying between Milton county on the southwest and Gwinnett on the northeast. A very limited amount of both vein and placer mining has been done on this portion of the belt.

Two or three miles from Sheltonville some shafts were sunk years ago on the Collins property and on the adjoining Adam Campbell property. Some drifting is reported as having been done at the latter locality and it is stated that a small stamp mill was erected when the mining was in progress. Nothing can be seen of the veins at the surface and the results of the work could not be ascertained. Some work was also done on the Settles and Strickland properties. For several miles along the course of Cowpens Branch considerable placer work was done. Sufficient placer gold in commercial quantities has been obtained in the Forsyth county portion of the Hall county belt to encourage vein prospecting.

Three or four miles from Sheltonville on lot 420, 1st district, Dr. E. D. Little formerly of Sheltonville, but now of Atlanta, sank some shafts a number of years ago and prospected several quartz veins. A milling test of some of the ore, made at the time, is said to have yielded satisfactory results.

GWINNETT COUNTY

The Hall county gold belt attains much greater commercial importance in Gwinnett county than in the counties to the southwest. Mining developments have, so far, been quite limited in extent, but enough has been done to prove the presence of some well defined quartz veins, yielding ore of a very good grade. The more important mining operations have been carried on in the vicinity of Buford, a town in the northeastern part of the county on the main division of the Southern Railway.

The gold belt traverses the northeast corner of the county and closely parallels the Southern Railway from which it is distant only about a mile on the northwest. The close proximity of this portion of the Hall county gold belt to stations on the main division of an extensive railway system, together with easy accessibility to Atlanta, the most commercially important city of the State, gives the Gwinnett county deposits unusual advantages of location.

The majority of the auriferous quartz veins examined were found to cut the trend of the inclosing formations at a distinct angle. In some of the veins, that have been worked, galena is a conspicuous sulphide. At the Piedmont mine it occurs in considerable quantities, and is reported at this locality as being argentiferous.

HARRIS PROPERTY.—This property, on lot 275, 7th district, is about a mile northeast of Suwanee. Two quartz veins occur here within something like a hundred yards of each other and prospect work consisting of shafts and open cut work has been carried on in the past at the locality. It is reported that one of these veins was mined before the Civil war, an arrastre being used for milling the ore. In Bulletin 4-A, of the Geological Survey of Georgia, published in 1896, it is stated that a sample taken from the southern vein and assayed by the Survey gave only a trace of gold.

Moore & Brogden Property.—Some auriferous quartz veins were prospected to a very limited extent several years ago at the above locality, lots 309, 310, 318 and 319, 7th district. Some panning tests of veins exposed in shallow pits, made by the Survey when Bulletin 4-A, on the Gold Deposits of the State was published in 1896, gave favorable results.

Suwanee, or Level Creek Mine.—This mine is about four miles west of Buford on lot 289, 7th district. The tract on which it is situated is known locally as the Simmons property. Mining operations were conducted at this locality and on the neighboring Shelly property before the Civil war. In later years, several different parties, or companies, have carried on limited mining operations on the Simmons lot. These more recent operations have been in the nature of prospect work rather than systematic mining. Enough, however, has been done in the course of combined operations to demonstrate the occurrence of a well defined quartz vein, and some ore of good grade has been obtained.

The vein on which the principal work has been done, and which may be designated as the "Simmons vein" to distinguish it from some contiguous veins of less importance, has been mined at the crest of a small hill and also a hundred yards or so to the east near a creek. The general strike of the vein, as indicated by an exposure in a cut on the hill and the line of workings, is about 12° north of east. While there is no direct evidence that the workings at the two localities just mentioned are on one and the same deposit, it is most probable that they are, judging from the strike of the vein. Near the creek a shaft was sunk many years ago which was known as the Simmons shaft, and in 1901 another shaft was sunk about two hundred and fifty yards

to the east of the Simmons shaft. At the time of a visit
to the property about five years ago the Level Creek Mining
Company was engaged in deepening this shaft and some
drifting from it was in progress. The last work done on the
property was on the hill previously mentioned. Commencing
at the edge of the creek lowlands an open cut was made by
hydraulic work and extended into the hill until the vein was
encountered. An inclined shaft had previously been sunk to
a depth of about a hundred feet near the crest of the hill
descending on the vein underneath the present open cut. A
few yards to the west from the mouth of the inclined shaft
another shaft has been sunk on the vein to the depth of about
sixty feet. A short distance from the latter shaft a well
timbered two compartment shaft was sunk in 1908 by the
Suwanee Gold Mining Company. It could not be ascertained
whether or not the vein was cut in this shaft. No mining
operations were in progress at time of visit and the underground works could not be entered.

In the open cut immediately below the mouth of the inclined shaft the vein was exposed for a distance of twelve or
fifteen feet and showed at this point a thickness of five or six
feet. A sample for assay was taken from this exposure and
also from an ore pile, of a number of tons on the top of the
hill, the ore of which was said to have been taken from both
the inclined and the sixty-foot shaft. These two samples on
assay yielded the following results:

> Sample from vein in open cut $7.44 per ton.
> Sample from ore pile _____ 4.15 per ton.

In Bulletin No. 4-A, of the Geological Survey of Georgia,
published in 1896, the results of an assay sample is given that
is stated to have been taken from a large ore dump near the
main pit. The exact location on the vein from which the sam-

ple was taken can not now be determined. The assay results showed a value of $16.44 per ton.

In addition to the Simmons vein, one or two other veins have been prospected to a small extent on the hill near the principal shafts. The exposures, however, were found to be so slight that nothing definite can be stated concerning the character of these deposits.

SHELLY PROPERTY.—This property, on lot 290, 7th district, lies adjacent to the tract of the Suwanee mine. At a point on the property known as Hay Mount Hill, considerable mining, consisting largely of open cut work, was carried on many years ago. A little prospect work was also done at a locality near this point by the Level Creek Mining Company when they were operating on the Simmons tract. At time of visit little, if any, exposure of the veins was observed. The property is regarded favorably by those familiar with the gold deposits of Gwinnett county.

In Bulletin No. 4-A, of the Geological Survey of Georgia, published in 1896, the results are given of an assay sample from a quartz vein on the Shelly tract. It is stated that the vein from which the sample was taken exhibited, where exposed, a maximum thickness of four and an average thickness of about two feet. The results of the assay showed a value of $23.50 per ton.

PIEDMONT MINE.—This mine, on lot 304, 7th district, is situated about two miles west of Buford. Considerable underground work, consisting principally of tunnels and shafts, was carried on here about fifteen years ago. A stamp mill was erected at the time the mine was being exploited and some of the ore milled.

No work has been done on the property in recent years and at time of visit the vein could not be inspected. As the mine was open to inspection when Bulletin No. 4-A, of the Geologi-

cal Survey of Georgia was published, the following description of the deposit is quoted from that report: "The Piedmont vein is plainly what is known among mining men as a
true fissure vein, cutting the country rock at a wide angle. It
strikes about due east, the strike of the country rock being
about N. 50° E. The ore body will average from eighteen
inches to two feet in width and consists of compact milk-white
quartz containing varying proportions of pyrite, galena and
free gold, the latter frequently apparent to the unaided eye.
Immediately adjoining the vein proper, or in its binding, are
small quartz stringers which will undoubtedly pay to work
with the main ore body. This ore body has been traced
already several hundred yards by test pits, and surface indications may be found along its lead for some distance."

Captain G. W. Thompson, of Buford, Ga., states that from
approximately eighty tons of ore, milled when the mine was in
operation, about $605.00 worth of gold was obtained. This
gentleman, who is in charge of the property, exhibited a portion of United States mint returns that had been preserved
showing receipts amounting to $451.49. The mine is owned
by the Piedmont Gold Mining Company.

PLACER DEPOSITS OF RICHLAND CREEK AND TRIBUTARIES.—
Considerable valley land on Richland Creek close to the Piedmont mine is reported as being auriferous. At the edge of
this tract, along a tributary branch, the Richland Gold Mining
Company worked an acre or two of placer deposits a number
of years ago. It was not learned what success attended these
operations. A little higher up the branch just mentioned considerable placer work has been done in past years. The strip
of valley land along the upper course of this branch is very
narrow, but it is reported that a good many thousands of dollars worth of gold was obtained. At the upper end of the old
workings it is claimed that a small vein occurs known as the

"Nugget" vein, numerous nuggets having been obtained at that point.

OWENS MINE.—The Owens mine, on lot 349, 7th district, is three miles west of Buford. About thirteen years ago some underground work was done here by Messrs. Liddell and Johnson, who erected at the time a rather extensively equipped stamp mill. Later a little work was done under an option by Mr. J. W. Green, of Buford. No mining operations have been carried on at the locality for some years and an examination failed to enable any conclusions to be drawn as to the character and value of the deposits. Captain G. W. Thompson, of Buford, Ga., is in charge of the property.

HALL COUNTY

The Hall county gold belt traverses the central part of Hall county in a southwest-northeast course. Gainesville, the county seat, situated on the main division of the Southern Railway, is on its southwest edge. The belt attains a maximum width in this county of six and a half or seven miles. Future investigations may demonstrate that some deposits in the neighborhood of Murrayville in the northern part of the county, described under the heading of "Isolated localities," should be included in the main belt.

In Hall, as in Gwinnett county, the mining operations carried on have been of a rather limited character, consisting in the majority of cases of prospect work rather than regular mining.

McCLESKY MINE.—The McClesky mine is about three miles west of Gainesville in the northern part of lot 5, 8th district. Mining on a small scale has been conducted here on an auriferous quartz vein at irregular intervals for many years. The mine was operated before the Civil war, by Mr. Green McCles-

key, by open cut work and flat inclines. After that period Mr.
Samps Mooney carried on mining at the locality for a number
of years erecting a stamp mill near by on McCleskey Creek.
About eight years ago, Messrs. Jaquish and Mathews carried
on systematic underground mining for a year and a half.
Under this management a very conveniently arranged and
well equipped milling plant was erected near the main shaft.
The mill was constructed for a ten-stamp battery, though only
five stamps were installed. A shaft was sunk on an incline to
a depth of about two hundred and forty feet and a drift driven
on the vein for approximately a hundred feet. About four
years ago the mine was operated for a short period by the
New York Development Company.

At time of visit the mine had been idle for over three years
and the underground works were inaccessible and no opportunity was afforded to inspect the vein. According to Mr. H.
D. Jaquish, as exposed in the work done by Messrs. Jaquish
and Mathews, it varied in thickness from a few inches to several feet.

BYRD MINE.—About a mile and a half northwest of Gainesville, on lot 156, 9th district, a very limited amount of mining
was prosecuted on an auriferous quartz vein a number of
years ago. At a point immediately adjacent to the Gainesville
public road a shaft, or pit, was sunk and it is stated by those
familiar with the deposit that considerable gold was obtained.

PROSPECTS IN THE VICINITY OF GAINESVILLE.—Prospect work
has been done in the past on several quartz veins near the
northeast edge of the city of Gainesville. No mining has been
carried on at any of the localities for a number of years and
nothing very satisfactory can be stated concerning the character of the deposits. On lot 138, 9th district, in the neighborhood of the stand pipe of the city water works, considerable

prospecting was done on a quartz vein occurring in a small ridge. Several tunnels were run into the hill to strike the vein, one passing entirely through the ridge. Mr. Twitty, of Gainesville, the owner of the property, states that at one time placer work was conducted along a small branch at the base of the ridge. About a mile from the Twitty prospect the Stowe mine was once the scene of active prospecting. A number of shafts were sunk on a quartz vein and a small milling plant erected and operated for a short period. Prospect work has also been done on the Longstreet property, lot 130, 9th district; on the Merk property, lot 129, 9th district; and on lot 127, 9th district. A few miles to the north of these localities some placer work has been conducted in the past along a creek emptying into the Chattahoochee River on the Guilford Thompson property.

THE IVY MOUNTAIN AND ELROD MINES.—These two properties are situated on adjoining lots, Nos. 103 and 104, 10th district, about six miles northwest of Gainesville on the public road to Dahlonega. Considerable work has been done in the past at both localities on vein deposits which are probably continuous on the two lots. Some large bodies of vein material occur here, but, where open to inspection, the appearance of the deposits would indicate a low grade ore. It is reported, however, that some rich pockets have been mined. In Bulletin No. 4-A, of the Geological Survey of Georgia, published in 1896, Mr. Francis P. King says of the deposits at this locality: "The schists of the region have a very slight dip and are apparently interbedded with seams of small milky-white quartz carrying a small percentage of sulphides. This seam varies in thickness from a few inches to several feet and is evidently a stringer vein horizontally inclined. The topography of the region is such that the vein is readily exposed by shallow pits and is to be found over a wide area. The major portion of this immense bed of quartz is of a very low grade ore, although in the past some rich pockets may have been found." On both properties a stamp mill was once erected and operated.

MAMMOTH VEIN.—Some prospect shafts have been sunk in the past on a large quartz vein occurring close to the Southern Railway and about three miles northeast of Gainesville. The vein is an unusually large and well defined one, but according to reports the ore is of a very low grade.

PASS PROPERTY.—Some prospect work was done years ago on lot 133, 10th district, about six miles north of Gainesville. A sample for assay was taken by the Survey, when Bulletin No. 4-A, of the Geological Survey was published, in 1896, from a vein exposed in an old shaft. The vein is stated as being sixteen inches in thickness where sampled. The sample yielded on assay $1.00 per ton.

CURRAHEE MINE.—The Currahee mine is situated on the Southern Railway about six miles northeast of Gainesville. Regular mining has been conducted in the past on an auriferous quartz vein a short distance west of the railway line and prospect work has also been done on a vein on the Thompson tract, a body of land lying on the opposite side of the railway. The last mentioned vein has been exposed in several shallow cuts or shafts; the deepest, about thirty-five feet in depth, situated in a small ravine, gives the best exposure of the vein. As seen here it has a northeast strike and a nearly vertical dip with an average thickness of about two and a half feet. This vein is quite clearly defined and consists of massive quartz containing considerable amounts of pyrite and galena. An assay, however, of a sample taken from an ore pile of a number of tons taken from the vein in sinking the shaft yielded only $0.40 of gold per ton. In addition to work on this vein some prospect work has been done on another vein on the Thompson tract.

Immediately west of the railway considerable mining has been done in past years on a vein known as vein No. 5. This vein has been exploited for several hundred yards by shafts, open cuts and tunnels. No work has been done on vein No. 5 for a number of years and little opportunity was afforded to draw conclusions concerning its character. A sample taken for assay from an ore pile at the mouth of one of the shafts, known as the Lilly shaft, yielded $5.79 per ton. A sample from an ore pile of a number of tons at the mouth of an open cut and tunnel near the old mill site and immediately adjacent to the Gainesville public road yielded no gold. Whether or not this ore was from vein No. 5 can not be definitely stated, but from the difference in direction between the tunnel and the strike of vein No. 5 it is probable that it came from a vein that is an offshoot from vein No. 5 or a cross vein. In appearance the ore here was almost identical with that of the vein on the Thompson tract previously described.

In addition to the Lilly shaft, above mentioned, two shafts, known as the Roberts and the Hayden shafts, were sunk to a depth of about two hundred and a hundred and forty feet respectively. These shafts located approximately sixty feet apart are stated to be connected by a drift at a hundred-foot level. About seven years ago the Currahee Mining and Milling Company erected a milling and cyanide plant at the mine. This plant was operated for a year and has since been destroyed by fire. It was not learned what success attended the operations.

In addition to vein No. 5, several other veins west of the railway have been prospected to a limited extent, but nothing definite can be stated concerning their probable value. At the time of visit the mine was controlled by Mr. Josephus Roberts, of Philadelphia, Penn.

GLADE MINE.—The Glade mine is situated eleven miles northeast of Gainesville and about a mile and a half northwest of the Chattahoochee River. Some prospect work has been carried on in the past on quartz veins at the locality, but the more important mining operations have consisted of work in placer deposits along several branches emptying into Flat Creek and hydraulic work in saprolite material.

In the northeast corner of lot 94, 12th district, some prospecting was done on a quartz vein a number of years ago by Mr. A. G. Jennings, the former owner of the property. A sample was taken for assay from an ore pile of a number of tons of vein quartz at the mouth of one of the old shafts. This sample yielded on assay only $0.20 per ton. Near by on lot 99 some old workings are to be found on another vein.

Considerable placer work was done before the Civil war on lots 116 and 117 along Stockeneter Branch. A portion of this placer deposit on lot 117 was re-worked by A. G. Jennings about 1880. Panning tests made from a fringe of unworked gravels near the edge of the extensive Flat Creek lowlands yielded very satisfactory results, the gold obtained being rather coarse. Whether there occurs along this branch much placer area that could be re-worked at a profit could only be determined by more extensive tests than it was practicable to make at the time of visit. Placer work has also been done on lot 100 along Glade Branch which empties into Flat Creek a short distance above Stockeneter Branch. Near Glade Branch a cut of considerable dimensions was made in saprolite material by hydraulic work. From a small colluvial gravel deposit at the top of this cut some panning tests yielded considerable amounts of gold. Placer work has also been done in the past on lot 99 along Deavours Branch on the opposite side of Flat Creek from the branches previously mentioned. Some placer mining in addition to that already described has been done in the lowlands along Flat Creek. These lowlands just mentioned are quite extensive and, as a number of branches that have afforded remunerative placer deposits are tributary to the creek at this locality, the advisability of operating a dredge boat in the main Flat Creek bottoms has been considered. Systematic tests would be necessary in order to determine whether gold is present in sufficient quantities to warrant this class of mining operations.

Only meagre records of the work done by A. G. Jennings at the Glade mine are now available. Mr. James Hunt, of Gainesville, the owner of the property, has, however, in his possession some old express receipts showing shipments of gold amounting, in some cases, to as much as a thousand dollars at a shipment.

The reported occurrence of diamonds, found when the gravel deposits along Stockeneter Branch were being washed, deserves passing mention. The report was given full credence at the time by Dr. M. F. Stephenson, who was the author of a small volume on the mineralogy and geology of Georgia.

HABERSHAM COUNTY

The Hall county gold belt traverses Habersham county in a southwest-northeast course. Clarkesville, the county seat, is within, or near, the belt. Systematic mining has been conducted in the Hall county belt in Habersham at only one locality, namely, the Nichols mine. The exact location of the belt is not as well known here as in the counties to the southwest.

PROPERTIES NEAR THE CHATTAHOOCHEE RIVER IN THE SOUTH-EAST PART OF THE COUNTY.—On lot 57, 10th district, Mr. James Crow prospected some quartz veins several years ago. The best exposure of any of these veins to be seen at time of visit was in a small shaft, or pit, about twenty feet deep. The vein showed a varying width in this shaft of from six inches to a foot and contained a moderate amount of iron pyrite and showed some free gold. Some decomposed walling from the south side of the vein also yielded gold when panned.

Near the Crow property, on lots 55 and 62, 10th district, a little placer work has been done along a small branch by Mr. Tillman Perkins. Mr. Perkins exhibited, at the time of visit, some of the gold obtained. A nugget weighing probably six or seven pennyweights was noticed.

NICHOLS MINE.—The Nichols mine is situated about six miles east of Clarkesville on lots 92 and 120, 12th district, and is close to the Tallulah Falls Railway. About five acres of a placer deposit occurring along a small branch have been worked for gold at different intervals in the past. It is said that a good many thousand pennyweights of gold have been obtained in the course of mining operations. Nuggets weighing over 200 pennyweights have been reported from this locality. No work has been in progress here for a number of years and the placer deposit has probably been pretty thoroughly worked over. A number of excavations have been made on the slopes adjacent to the placer works in efforts to locate auriferous veins. Several quartz veins were exposed by this work but none of them seem to have warranted permanent mining. Some of the quartz veins exposed exhibited comb structure. Here, as at a number of placer deposits in the State where good values have been obtained, more thorough and systematic prospecting might locate valuable vein deposits.

RABUN COUNTY

The Hall county gold belt traverses the southern portion of Rabun county in a southwest-northeast direction. It closely parallels in its course the Chattooga River, lying a short distance to the northwest of that stream. Mining operations on the Hall county belt in Rabun county have been confined, with the exception of a little prospect work, to placer deposits at three or four localities.

BRIGHT EVANS PROPERTY.—This property is situated in the extreme southeastern part of the county near Tallulah River. Some prospect work was done on a quartz vein at this locality at the base of Long Mountain a number of years ago. The vein is described in Bulletin No. 4-A of the Geological Survey of Georgia, published in 1896, as being exposed at that time in a small open cut and having a thickness of from ten to eighteen inches and showing some free gold. A sample of the ore assayed by the Survey, however, yielded only a trace of gold.

LAMAR MINE.—The Lamar mine, on lot 30, 2d district, is a placer deposit occurring along a small branch tributary to Warwoman Creek. The locality is about ten miles east of Clayton, the county seat of Rabun county. The principal mining operations carried on here were done before the Civil war. The bed of the stream and a considerable part of a body of narrow valley land bordering the branch have been worked for a distance of a half a mile or more. Some panning tests made from unworked portions of the gravels yielded fair amounts of gold.

HAMBY PLACER.—A small stream known as Hamby Branch has been worked for gold years ago. This placer, on lot 43, 3d district, lies immediately north of the Lamar mine.

PAGE PROPERTY.—This property, lots 44 and 45, 3d district, lies a few miles to the northeast of the Lamar mine on the

Chattooga River near the mouth of Laurel Creek. A considerable amount of placer mining has been done along Page's Creek and Lawground Branch. From the placer deposits along Lawground Branch it is reported that a good many nuggets have been found weighing from five to twenty pennyweights.

HEDDEN PLACER MINE.—This mine is located in the extreme northeast corner of the county on lot 100, 3d district, about three and a half miles north of Pine Mountain post office. The Hedden mine is a placer deposit along Hedden Branch and a small tributary. The principal mining operations conducted here were carried on years ago; the greater portion of the work was done on lot 100. It is reported that considerable amounts of gold were secured, some portions of the deposits having been worked several times. The most productive areas have probably been pretty well exhausted. A little prospect work for veins had been in progress at the locality shortly before the time of visit. Nothing was noticed, however, in the pits examined that looked very promising for a future supply of ore.

Northeast of the Hedden mine the Hall county gold belt extends into Macon county, North Carolina. The Ammons Branch mine in Horse Cove in that county is probably on the Hall county belt.

THE CARROLL COUNTY BELT

CARROLL COUNTY

The Carroll county belt extends in this county from a point near the southwest corner to the northeast corner. It passes in its course close to Carrollton, the county seat, and through the town of old Villa Rica, situated about a mile to the west of the modern town of that name on the Birmingham Division of the Southern Railway. Southwest of Carrollton, systematic mining has been pursued at only one locality, namely, the Bonner mine. In the vicinity of Villa Rica, however, considerable mining has been carried on at a number of points. According to reports, a large amount of gold was obtained at the Villa Rica locality in the early mining days from surface washing, the town having received its present name, signifying rich village, from the results of these mining operations.

BONNER MINE.—This mine, on lots 94 and 99, 11th district, is situated about seven miles southwest of Carrollton and close to Buffalo Creek. The most important mining operations have been confined to lot 99. Gold was discovered here in the early forties in some branches tributary to Buffalo Creek. It is stated that something like a half a million pennyweights of gold were secured before the Civil war from the mining at this locality of placers along the branches and surface deposits of dry hollows and slopes adjacent to the placers. In later years, vein mining has been conducted on the property.

The vein deposits that have attracted attention are located in a ridge having an elevation of about one hundred and fifty feet above Buffalo Creek which flows at the base of its northern slope. The deposits may be described as a more or less continuous zone of varying thickness composed of quartz stringers and interlaminated gneiss or mica schist. This zone has been prospected for more than a mile along the ridge. It is stated that at some points the thickness of the ore bodies is as much as forty feet. The best exposure is to be found in a large shallow cut several hundred feet long near the Bonner residence. The results of seven assays of samples taken by the Survey from the Bonner mine and given in Bulletin No. 4-A, of the Geological Survey of Georgia, published in 1896, are here quoted:

No. 1. Ore sample from Big Cut, Bonner Mine
                      .05 oz. ($1.00) of gold per ton.
No. 2. Ore sample from the same cut
                      .05 oz. ($1.00) of gold per ton.
No. 3. Ore sample from Bradley Pit, Bonner Mine
                      .05 oz. ($1.00) of gold per ton.
No. 4. Ore sample from the same pit
                      .05 oz. ($1.00) of gold per ton.
No. 5. Ore sample from the Tuttle Shaft
                      .125 oz. ($2.50) of gold per ton.
No. 6. Ore sample from the same shaft
                      .250 oz. ($5.00) of gold per ton.
No. 7. Ore sample from Bonner Mine
                      .14 oz. ($2.80) of gold per ton.

A milling plant is located on a branch a short distance from the large cut.

A considerable tract of valley land occurs along Buffalo Creek that is thought to contain valuable placer deposits, but owing to heavy overburden and slight drainage fall very little mining has ever been attempted.

The presence of garnets in considerable quantities in the gneiss at the Bonner mine is worthy of note. A number of lenses of coarsely banded, highly feldspathic gneiss were noticed in intimate association with the ore bodies. Some of these may represent true pegmatite dikes. The absence, as far as observed, of any rocks of a basic type, together with the structural relations, would suggest an original acidic rock magma as the probable source of the gold at this locality.

STACY PROSPECT.—The Stacy property is situated on Oak Mountain about four miles east of Carrollton. The location would seem to be three or four miles east of the general trend of the Carroll county belt, but, excepting this property, no mining has been done northeastward from the Bonner mine along the belt for a distance of about twelve miles and exact boundaries in this section have not been definitely located.

The work done at the Stacy place consisted of prospect work on a quartz vein said to be about three feet thick and consisting of granular or saccharoidal quartz. At the time mining operations were carried on a small stamp mill was erected and some of the ore milled. No work has been done at the locality for a number of years and nothing definite can be stated concerning the character of the deposits.

J. L. Davis Property.—This property, located about four miles southwest of Villa Rica, has been prospected for gold to a limited extent.

Hixon Property.—Some prospect work was done a number of years ago on a quartz vein occurring in a ridge on this property, lot 166, 6th district. The locality is about two miles west of Villa Rica. A well defined quartz vein is to be found here, but the ore examined showed few sulphides and its appearance was not promising looking.

Hart Mine.—This mine, on lot 165, 6th district, is about a fourth of a mile northeast of the Hixon property. A shallow open cut about seventy-five yards long was made years ago on a quartz vein at this locality. Later, a shaft was sunk on the vein from the bottom of the old cut. No satisfactory conclusions as to the character of the vein could be drawn from an examination of the old works. According to reports, some ore of a good grade was obtained when mining operations were in progress.

Lassetter Property.—This property, lot 189, 6th district, is about a fourth of a mile from the Hart mine. Several shallow shafts, or pits, have been sunk on a large quartz vein occurring along the crest of a small ridge at this locality. Few sulphides were noticed in the vein quartz, and the ore examined did not look promising.

JONES MINE.—The Jones mine is about a mile northeast of the last described properties. The mining that has been done here was carried on before the Civil war under the direction of Mr. Jack Jones, formerly Treasurer of Georgia. According to report, an open cut about a hundred yards in length was made on a quartz vein and a shaft sunk and considerable drifting done. The old works have almost filled up and nothing could be seen of the vein. It is stated by those familiar with the history of the mine that good values were obtained from the ore milled.

In addition to the vein mining, considerable surface washing was done on a slope near the vein and a placer deposit was worked along a branch contiguous to the mine. The property is owned by Messrs. Oscar Fielder, Will Hamrick and others of Villa Rica, Ga.

CHAMBERS MINE.—This mine, on lot 195, 6th district, adjoins the Jones tract and is about a mile northwest of Villa Rica. Considerable mining has been prosecuted on an auriferous quartz vein at this locality and surface washing, covering an area of a number of acres, has also been done. An open cut was made before the Civil war on the vein referred to and was enlarged in the course of more recent mining operations. Some shafts were also sunk on the vein. The cut is about a hundred and fifty yards long. At one point, where a small portion of ground was left standing for a passage way a limited exposure of the vein is to be seen. At this point it shows a thickness of about four feet. No work has been done on the vein for a number of years and it could not be ascertained what gold was obtained when mining operations were being carried on.

A limited amount of work was done near this vein in 1908. These operations consisted in surface washing near the northwest side of the vein. The material was carried in a flume across the open cut and washed by machinery situated on the southeast side. Work had been suspended before the property was visited and it was not learned what success attended this mining. Near the southwest end of the cut surface washing was done years ago over an area of a number of acres. According to those who have worked at the locality, a number of small comparatively rich pockets have been found in the vicinity of the main vein. The gold in these is said to have been in very thin quartz veins, or quartzose layers in the decomposed rock.

CLOMPTON MINE.—This mine, on lot 194, 6th district, is a short distance to the northeast of the Chambers mine. Considerable work was done here a number of years ago on an auriferous quartz vein. The work consisted of shafts, open-cut work and tunnels. According to Mr. Clarke Watkins, of Villa Rica, who is quite familiar with the history of most of the gold mines of this portion of the Carroll county belt, about thirteen hundred pennyweights of gold were obtained at this locality from a cut of comparatively insignificant size. At the time the principal mining operations were being carried on a milling plant was installed. The millhouse is still on the property, but the machinery has been removed.

ASTINOL PROPERTY.—This property is close to the Clompton mine. Considerable prospect work was done here a number of years ago. No mining has been in progress in recent years and it could not be learned what success attended the former operations.

SOUTHERN KLONDYKE MINE.—This mine is situated about a half a mile to the northeast of the last two described properties. Considerable mining was conducted at this locality some years ago on an auriferous quartz vein. Owing to lack of surface exposures, and the fact that the underground works could not be entered at the time of visit, nothing definite can be stated concerning the character of the deposit. The mine is equipped with a conveniently located milling plant situated on a creek near by. The mill is a stamp mill of the ordinary type and there are a number of Frue vanners.

DOUGLAS COUNTY

The Carroll county gold belt passes through the extreme northwest corner of Douglas county, its total extent here being only a few miles. The nature of the deposits and the character of mining operations that have been carried on differ in no important particulars from those of the adjacent Carroll county portion of the belt.

PINE MOUNTAIN PROPERTY.—Pine Mountain, on lot 206, 2d district, is a hill rising something like two hundred feet above the neighboring valleys. The locality is about three miles northeast of Villa Rica. Considerable mining has been done here at different times. The operations have consisted of both surface washing on the slopes of the ridge and also of shafts and open-cut work on quartz veins. Some very large bodies of quartz occur along the crest of the ridge, but, in the opinion of persons most familiar with the deposits, the large quartz veins occurring here are too low grade to work with profit, the paying values being confined to small stringers and lenses of quartz. Considerable work, however, seems to have been done at one time on the northwest side of the mountain, a short distance beyond a slight depression on several veins of some magnitude known as the Stubblefield veins. It was not learned what success attended these operations. The advisability of washing the saprolite material on the slopes of the mountain on a large scale has been considered. Water could be obtained by pumping from a creek about three-fourths of a mile distant. Gold is probably widely distributed over a considerable area at this locality, but if hydraulic mining on an extensive scale was undertaken a large amount of material carrying little or no values would probably have to be moved. From a few panning tests made from the saprolites a small amount of rather fine gold was obtained. Opportunity was not afforded, however, to make any extensive tests.

McManus Property.—This property, which is a short distance northeast of Pine Mountain, has been prospected for gold. No work has been done recently and nothing definite can be stated concerning the character of the vein.

Mine No. 212.—This mine, on lot 212, 2d district, is about a half of a mile to the northeast of Pine Mountain. Some mining was carried on at this locality before the Civil war. The most important operations, however, were conducted here about sixteen years ago by a company composed chiefly of English capitalists. Several shafts were sunk and considerable drifting done. The two most important shafts are said to be something over a hundred feet in depth and the drifts to have an aggregate length of several hundred feet. A stamp mill and other machinery were installed on the property at the time the mining was being carried on. As the old works were inaccessible, nothing definite can be stated concerning the character of the vein.

Triglone Mine.—This mine, on lot 204, 2d district, is about a half a mile from Mine No. 212. Open-cut work was done here on an auriferous quartz vein before the Civil war by Thomas Willoughby and Robert Triglone. About sixteen years ago considerable underground mining was done by a company who worked the property for a couple of years. The main shaft is said to be about two hundred feet deep and a good deal of drifting was done. Owing to the vein having been taken out at the surface in the old open cut work above mentioned. no exposures were noticed and the underground works were inaccessible at time of visit.

PLACER DEPOSITS ON LOT 77.—Placer mining was conducted in the early mining days along a creek and a tributary branch on lot 77, 2d district. Work has been done on the Henslee, the Duncan and one or two other adjoining properties.

PAULDING COUNTY

The Carroll county gold belt traverses the southeastern portion of Paulding county. Very little mining has been done on the Paulding county portion of the belt, though to the northeast, in Cobb county, work has been carried on at a number of localities.

AUSTIN MINES.—The Austin mines are old placer works along a small stream on lot 984, 2d district. The location is about seven miles southeast of Dallas.

Northeast from the Austin mines no regular mining for gold seems to have been carried on until the Mason mine is encountered in Cobb county. It is reported, however, that gold occurs at the Paulding Mining Company's pyrite mine, which is situated near the eastern edge of the county and on the trend of the gold belt.

COBB COUNTY

The Carroll county gold belt traverses the northeast corner of Cobb county, passing in its course a few miles to the northwest of Lost Mountain and crossing the Western and

Atlantic Railway about a mile and a half southeast of the town of Acworth. Two or three miles northeast of the last mentioned locality it unites with the Dahlonega gold belt near the Cherokee-Cobb county line.

MASON MINE.—The Mason mine is situated on lot 342, 20th district, and is about eight miles northeast of Dallas, the county seat of Paulding county. An auriferous quartz vein was located here about ten years ago, and in 1906 a stamp mill, and shortly afterwards a small cyanide plant, were erected and regular mining operations inaugurated. When visited, the mine was closed down and very little could be learned of the character of the vein from the limited exposures noticed. It was stated that at the time of examination the main shaft was a hundred and thirty feet in depth and that considerable drifting had been done. A small exposure of the vein was seen near the main shaft in a tramway excavation and also in a prospect pit about fifty yards to the northeast. The vein where exposed showed a thickness of several feet, but contained some interlaminated wall rock. The underground works being inaccessible at the time of visit, no samples for assay were secured.

The milling and cyaniding plant is situated on a small branch a short distance from the mine. The ore is crushed in a stamp mill with a ten-stamp battery and treated directly with the cyaniding solution, no effort being made to save part of the values by amalgamation. The property is owned by Messrs. G. W. Mason and T. P. Cochrane.

HATHAWAY PROPERTY.—Some prospect work was done several years ago on the Henry Hathaway place a few miles northeast of the Mason mine. In Bulletin No. 4-A, of the Geological Survey of Georgia published in 1896, State Geologist, S. W. McCallie, gives the following description of the ore body as seen in several shallow open cuts: ''The vein material, which varies from two to four feet in thickness, consists of thoroughly decomposed mica schist with numerous thin layers of quartz. The latter frequently show free gold on their surfaces where they come in contact with the schist. A number of pan tests made at the different openings gave very satisfactory results. The dip of the vein, and also of the country rock, is to the southeast at a high angle while the strike is northeast and southwest.''

J. B. KEMP PROPERTY.—This property joins the last described property. Considerable prospect work has been done here at different times. Shallow shafts, or pits, have been sunk at several points, but at the time of visit little opportunity was afforded to inspect the ore bodies. A few years ago a shaft was sunk to a depth of about fifty feet and a small mill erected and operated for a short period. In Bulletin No. 4-A, of the Geological Survey of Georgia published in 1896, the results of four assays are given that were taken from an excavation some distance to the northeast of the point where the last mentioned mining was done. It is stated that the vein where sampled showed a thickness of about a foot.

The results of these assays were as follows:

No. 1. Ore sample, Kemp property,
.125 oz. ($2.50) of gold per ton.

No. 2. Ore sample, Kemp property,
.250 oz. ($5.00) of gold per ton.

No. 3. Ore sample, Kemp property,
.375 oz. ($7.50) of gold per ton.

No. 4. Ore sample, Kemp property,
.350 oz. ($7.00) of gold per ton.

Northeast of the Kemp place no regular mining, as far as could be learned, has been done along the belt until the properties mentioned below are encountered about seven miles distant and close to the Western and Atlantic Railway.

PAYNE, KENDRICK, RANDALL AND HOUSE PROPERTIES.—These properties are situated about a mile and a half southeast of Acworth and only a short distance from the Western and Atlantic Railway. Gold was discovered at this locality in placer deposits along branches tributary to Proctors Creek and these placers, covering comparatively small areas, were worked before the Civil war. Old underground works are also to be found adjacent to the placers, but nothing definite can be stated concerning the character of the veins.

FREEMAN MINE.—This mine, near the last described properties, is about a mile and a half southeast of Acworth and a half a mile west of the Western and Atlantic Railway. Regular mining operations were conducted on an auriferous quartz vein at this locality a few years ago. At the time of visit the underground works were inaccessible and no opportunity was afforded to inspect the vein. According to report some of the ore obtained when mining was in progress was quite rich. Several specimens of vein quartz were found on the old dump that showed free gold. It is stated that the main shaft is about two hundred and thirty feet deep and that about a hundred and forty feet of drifting has been done. It was regretted that no opportunity was afforded to examine the vein at this mine, as parties familiar with ore bodies give favorable opinions of their value. Mr. D. E. Maxwell, of Jacksonville, Fla., is reported as having control of the property.

HAMILTON MINE.—The Hamilton mine is a short distance to the northeast of the Freeman mine and on the eastern side of the Western and Atlantic Railway. Some vein mining was done here a number of years ago. An examination of the old works threw little light on the probable value of the ore deposits, neither could anything definite be ascertained as to what success attended the mining that had been carried on.

Northeast from the Hamilton mine, which is only a short distance from the Cherokee county line, the Carroll county belt merges into the Dahlonega gold belt.

THE DAHLONEGA BELT

HARALSON COUNTY

ROYAL MINE.—This property is the only one on which any extensive mining for gold has been prosecuted in Haralson county. The Royal mine is located near Walker's Creek, lot 134, 8th district, about three miles southeast of the town of Tallapoosa. Gold was discovered and both vein and surface mining carried on at this locality as far back as the early forties by William Owens, the original owner of the property. This mine was formerly known as the Hollins mine. It was purchased in 1896 by the Royal Gold Mining Company, who gave it its present name. The property is reported at the present time as being in the hands of the Haralson Mining Company, Mr. John R. Miller, of Redding, Penn., being president.

Considerable surface work has been done at different times in the immediate vicinity of the mine. From one of these old excavations, extending for about two hundred yards along the strike of the vein, it is claimed that over one hundred thousand pennyweights of gold were secured. The mine has not been in operation for several years and at the time the property was visited the underground workings were inaccessible and the vein could not be examined. The ore body is said to be composed of intercalated quartz stringers and schist forming a zone several hundred yards in thickness and dipping steeply with the country rock.

One of the most extensively equipped plants for treating refractory gold ores, to be found in the State, is located at this mine. A commodious mill house containing four ten-stamp batteries, concentrating tables, etc., is located near the main shaft, in addition to a smaller adjoining mill house supplied with a Huntington mill. A large roasting and chlorination plant is situated convenient to the milling plant.

EDWARDS MINE.—Some placer work on a small scale has been carried on quite recently on lots 63 and 64, 8th district. A limited area of the gravel deposits along a small stream were exploited in the course of these operations. Work had been suspended before the property was visited and it was not learned what gold was obtained.

On the adjoining property of Mr. J. W. Harris, a thin stratum of subangular gravel was found exposed in a small gulch leading to the Edwards Mine. A little gold was panned from the gravel and Mr. Harris had in his possession a number of particles of rather coarse gold that his son had picked up from the bottom of the gulch after showers of rain.

OTHER PROPERTIES IN HARALSON COUNTY.—A little placer work has been done in the past along Pole Branch about three-fourths of a mile north of Buchanan, the county seat of Haralson county. Northeast of this locality some placer work was also done at one time along a small stream on the Dean property, lot 23, 7th district, and also on some nearby lots in the 20th district.

PAULDING COUNTY

The principal developments along the Dahlonega belt in Paulding county have been confined to the neighborhood of

Yorkville near the Haralson county line, and the region about Huntsville on Burnt Hickory ridge northeast of Yorkville. With the exception of very limited prospect work, nothing has been done on the Dahlonega belt in this county for over ten years. Little, therefore, can be said at present in describing the majority of the properties further than to give their location.

YORKVILLE MINE.—This mine is on lot 331, 19th district, and is about two and a half miles east of Yorkville. Gold was discovered here and a placer deposit along a branch worked about 1855. This placer is reported to have yielded good returns and gold-bearing veins were discovered on a hillside at its upper end in 1869. Vein mining was conducted on a small scale for a number of years after this discovery. About 1894 a Chicago company commenced work with the intention of operating the mine on an extended scale. A tunnel was driven on the vein for 450 feet along its strike and a number of cross cuts made. For some reason, this company failed in putting the mine on a permanent working basis. The rocords of Paulding county show that in 1898 the property was sold to J. W. Spicer, J. B. Book and Jerome Crane.

As the mine was in a much better condition for examination, when described by State Geologist, S. W. McCallie, in Bulletin No 4-A of the Geological Survey of Georgia, published in 1896, than it is at present, his description of the veins is here quoted. ''There are exposed on the hillside, in the several excavations, three different auriferous veins, all running parallel and dipping with the country rock to the southeast. Two of the veins seem to be a series of ore chutes rather than well defined continuous veins, while the other, which is now being worked, continues for some distance with an average width of fifteen feet. The ore bodies consist of a dark colored mica schist with quartz, all more or less contorted and impregnated with auriferous pyrite. The weathered surface of the vein is usually quite rough and iron-stained, and frequently shows free gold.''

OTHER LOTS NEAR THE YORKVILLE MINE.—Prospect work, consisting of a number of shafts, short tunnels and some open cut work, was done on lots 334 and 369, 19th district a number of years ago.

Some prospecting has also been done on lots 366, 488, 694 and 696 in the 19th district.

SCHOFIELD MINE.—Between the Yorktown portion of the belt and the Huntsville region, a little hydraulic work was done a year or so ago on the property of Mr. Oliver Finch. A small cut was made in saprolite material. The property was visited, but little could be learned regarding the nature of the deposit and no information was obtained as to what gold was secured.

About two miles southwest of Huntsville, placer deposits on a small tributary of Raccoon Creek have been worked at different times. A company, known as the Michigan Gold Mining Company, operated a small hydraulic plant at the locality about 1895.

SHEFFIELD PROPERTY.—Considerable prospect work was done a good many years ago near Huntsville on lot 656, 3rd district. It is reported that float ore, showing free gold has frequently been found on the property. Some ore was milled when prospect work was in progress, but no regular mining was attempted.

DUNAWAY MINE.—This mine is northeast of the last described properties near the Cartersville road and about three miles distant from Huntsville. The mineral interest of the property is owned by Mr. John M. Sears, of Thomkinsville, N. Y. Mining on a small scale has been carried on here at irregular periods for a number of years. The deposit which has attracted most attention occurs along the contact of a mica schist and a rock that on weathering yields residual material of a deep yellowish brown color. Only highly decomposed specimens could be obtained, but it is doubtless a hornblende schist, a rock identifiable at several points on Burnt Hickory Ridge, and belonging to the class of rocks rich in hornblende, characteristic of the Dahlonega belt between the locality just mentioned and Macon county, N. C. Ore material was not exposed in sufficient extent to enable the geological relations to be satisfactorily studied. A number of small quartz stringers were observed at one or two points scattered through decomposed rock the whole forming a stratum several yards in thickness. The character of the deposit is probably that of a small gold-bearing zone along the contact of the two rocks. The strike of the zone is northeast and southwest with a southeasterly dip of approximately 50°.

A shallow open cut about fifty yards long and twenty wide has been made along the zone in the course of mining operations. Pannings of the saprolite material at several points in this cut yielded fair amounts of gold, all of it being quite fine.

A stamp mill of small capacity is located on a branch near the open cut. No work was in progress when the mine was examined, but it is reported that good values were obtained from the material milled.

As the chances of the zone continuing for some distance along its strike seem favorable, and as the ore could be mined quite cheaply to probably a considerable depth in the decomposed rock, it is hoped that further developments will take place.

PLATE V

PLACER MINING AT THE COOSA CREEK GOLD MINE, NEAR BLAIRSVILLE, UNION COUNTY, GEORGIA

OTHER PROPERTIES ON THE DAHLONEGA BELT IN PAULDING COUNTY.—Northeast from the Dunnaway mine, only a little prospect work is to be noted along the belt until the neighborhood of Allatoona in Bartow county is reached. In addition to the properties described, small placer deposits have been reported at a number of points along Raccoon Creek and its tributaries. One of these, on lot 108, 3rd district, about four miles from McPherson, received considerable attention a number of years ago. •

BARTOW COUNTY

The Dahlonega belt passes through a small area of the extreme southeast corner of Bartow county. The section between Allatoona, on the Western & Atlantic Railroad, and Cherokee county, was once the centre of considerable activity in both vein and placer mining. For a number of years, however, practically nothing has been done on the properties of this region. This is probably due, in part, to the iron, manganese, ocher and other valuable mineral deposits of Bartow county near by claiming the attention of those interested in mining developments.

McDANIEL PROSPECT.—A gold-bearing vein, on lot 1,075, 21st district, was prospected for a distance of several hundred yards a number of years ago by the late Mr. McDaniel, for several years station agent for the Western & Atlantic Railroad at Allatoona. The work consisted of several small open cuts and pits along the strike of the vein. The ore is reported as being of low grade.

ALLATOONA VEIN.—Considerable work was done many years ago on a vein of the above name, on lot 929, 21st district. At one point the vein is said to have been mined and milled down to water level for a distance along its strike of something like two hundred yards. The vein is reported as having a varying thickness of from one to four feet. A small branch known as Gold Branch, which enters Allatoona Creek above the Allatoona vein lot, has old placer works along its course for about a half mile. Much of this placer was worked over a number of times and it is said to have yielded large amounts of gold.

GLADE MINE.—Considerable work has been done at different times in the past on quartz veins on several lots in the 21st district about two miles northwest of Allatoona. Several shafts are to be seen on lots 878 and 879. The two principal ones are known as the Eastport and Francisco shafts. Neither could be entered when the property was examined and nothing can be said regarding the character of the veins. The last work done at this mine was by an English company, whose operations were at the two shafts above mentioned. The property is now owned by Mr. D. A. Smith of Allatoona, Ga.

OTHER PROPERTIES IN BARTOW COUNTY.—In addition to the localities already mentioned, prospecting has been done on lots 974, 808, 1,097, 1,224, and an adjoining lot, all in the 21st district.

CHEROKEE COUNTY

The Dahlonega belt, entering Cherokee county from the corners of Bartow and Cobb, widens considerably and has proved, up to the present date, much more commercially important in this county than it has in any of the southwestern counties. While the placers have received the usual amount of attention, underground work has been more vigorously prospected here than at most localities on the belt. The Creighton mine, located in Cherokee county, with its extensive underground developments and successful treatment of refractory sulphide ores through a long period of years has undoubtedly encouraged systematic vein mining in Georgia more than any other one mine on the Dahlonega belt.

GEORGIANA MINE.—This mine is located on lot 958, 21st district, three and one-half miles northeast of Acworth in Cobb county. No work has been done at the Georgiana mine for a number of years. As an examination of the property in its present condition was found to be very unsatisfactory, the following description of the veins as given by State Geologist McCallie in Bulletin No. 4-A, of the Georgia Geological Survey, published in 1896, when the property was in better condition for inspection, is here quoted. "A number of auriferous veins have been located on this property, and it has been more or less prospected in the last few years. These veins occur along the side of a ridge near the headwaters of Fox Creek, where they are found to conform in strike and dip to the country rock. The prospecting consists of several tunnels and shafts, located at various points on the ridge. The main tunnel, four hundred feet in length, driven at right angles to the strike of the schists, cuts three different ore bodies which vary from two to twenty-seven feet in width. They consist principally of mica schist with but little quartz. The ore is all low grade, and thus far it has not been milled with profit. A second tunnel, one hundred and forty feet long, exposes an eight-inch vein carrying ore that is said to have yielded $5.00 per ton. The three principal shafts, all of which were inaccessible at the time of our visit, attain a depth of from ninety to one hundred and twenty feet. One of these cuts a quartz vein from three to six feet wide carrying gold in paying quantities. From these various excavations, several tons of ore are reported to have been taken from time to time and milled with satisfactory results."

With the mine in its present condition, no attempt was made to secure samples for assay.

GRANVILLE MINE.—About two miles from the Georgiana property, some vein mining has recently been done by Messrs. A. Granville and L. P. Roquette.

A shaft was sunk to a depth of seventy-five feet and a drift driven for about one hundred and eight feet on an auriferous quartz vein. The mine was not in operation at the time it was visited and the shaft could not be entered, and nothing can be stated concerning the character of the vein. About one hundred yards distant from the above described works an inclined shaft has been sunk to an approximate depth of thirty feet on a large quartz vein. Some ore examined here was not very promising looking, consisting of milky white quartz and showing little sulphides. A small test stamp mill has been installed. The shafts are well timbered, and the mining that has been carried on seems to have been done in a neat, systematic way.

WILLIAMSON PROPERTY.—An area of several acres of placer has been worked in the past on this property, lot 1,120, 21st district, and several veins prospected. A nugget, weighing eighteen and one-half pennyweights, that was obtained from the property some years back was shown the writer when the locality was visited. No mining operations have been carried on recently. Bulletin No. 4-A, of the Geological Survey of Georgia, records the results of two assays of ore from this property, each yielding $7.50 per ton.

KITCHENS PROPERTY.—Prospect work on gold-bearing veins has been carried on in the past on lots 823 and 834, 21st district. As an examination of the veins could be made to better advantage at the time Bulletin No. 4-A, of the Geological Survey of Georgia was published, a description, in part, of a large vein that has attracted the most attention is here given. "The so-called Kitchen vein, the large vein mentioned above, about which there has been lately much talk, occurs on lot 823, a short distance northwest of the Kitchen residence. It outcrops here along the side and near the top of a low ridge where it has been prospected at a number of places by open cuts from two to eight feet in depth. The auriferous ore body, which has been improperly called a fissure vein, consists of a zone or belt of garnetiferous mica schist, interlaminated with innumerable layers of quartz, all dipping at a high angle to the southwest."

The results of two assays of samples from this vein, given in the publication just quoted from, show values of $2.50 and $7.50 per ton.

Kellogg Mine.—This mine is located on lot 1,113, 21st district. Considerable placer work has been done in the past along a branch running through the lot and a number of shafts and pits have been sunk on different veins. The most productive portions of the placer deposit have long since been worked out, but the veins have only been prospected. Mr. George Hesselmeyer, who resides on the property, has the mine in charge.

South of Mr. Hesselmeyer's residence, a number of old shafts and excavations are to be seen that were made before the Civil war. Traces of an old water ditch about a quarter of a mile long are still visible. Nothing can be seen of the veins at this locality. To the north on the opposite side of the dwelling house, several shallow pits have been sunk at different points on a quartz vein known as the Sandstone vein. A sample for assay was taken from a section across this vein where it was exposed in a pit about fifteen feet deep. This sample yielded no gold. Bulletin No. 4-A, of the Geological Survey of Georgia, shows the results of an assay of a sample taken from the same vein as $2.50 per ton. The vein, as exposed in several pits along its strike, is not of uniform width, but occurs in bunches or kidneys.

About five years ago a shaft was sunk at another locality not far from Mr. Hesselmeyer's house to a depth of ninety-eight feet. It is stated that it cuts one or more veins. No drifting was done from this shaft. Some surface sluicing was done near this shaft.

A Lane Improved Chilian mill, with a Wilfley concentrating table and a twenty-five horse-power engine, all in a good state of preservation, are now on the property. In addition to this machinery, there is a log washer that was installed in the course of some experimental work with surface material.

SOUTHERN STAR MINING COMPANY'S PROPERTY.—Considerable prospecting for gold has been done at different times on lot 901, 21st district, formerly known as the Cox property. The Southern Star Mining Company is at present operating an iron pyrite mine at the locality, the underground works of which connect with works cutting an auriferous vein. The vein could not be inspected at the time the property was visited.

ADJOINING PROPERTIES TO THE SOUTHERN STAR MINING COMPANY'S PROPERTY.—Several veins have been prospected on the Bell property, lot 900, 21st district, and on the Evans or Cobb property, lots 792 and 793, 15th district. These lots lie northeast of the last described property. On the Bell property, considerable open cut work was done at one time on a quartz vein of a granular texture and more or less laminated, known as the Sandstone vein. A vein of the same character is found on the Evans or Cobb lots. Placer deposits were also once worked on the last mentioned lots along Rose Creek. A stamp mill was in operation at one time on both properties, but, with the exception of a little prospect work, nothing has been done at either locality for a number of years.

BENTLEY PROSPECT.—Some prospect work has lately been done on lot 104, 21st district, owned by Mr. J. J. Bentley of Acworth, Ga. Several narrow bands of siliceous material, carrying some iron pyrite and interlaminated with the country rock, are to be seen in the bed of a small branch on this property. These bands, where exposed, vary from two to eight or ten inches in thickness. Judging from the limited exposures, they appear to be impregnations of the country rock rather than true veins. A shaft about twenty feet deep has been sunk close to the branch on one of them. Water prevented the securing of an assay sample from this shaft. Two samples, however, were assayed that were taken from two of the bands a few rods higher up the branch, but no gold was obtained. Mr. Bentley later furnished some ore that he stated he had secured from the bottom of the shaft. This, on assay, yielded $3.31 per ton.

CHEROKEE MINE.—This mine is located on lot 428, 15th district, a few miles west of Holly Springs, a station on the Marietta and Knoxville Division of the Louisville & Nashville Railroad. The property is rather widely known, and has been worked for gold at different times for over fifty years. About ten years ago it was the scene of extensive mining operations, but no work has been done recently.

Placer deposits occur at the base of a prominent ridge in which a number of gold-bearing quartz veins are found. These placer deposits, having long since been worked out, attract no attention at present. Messrs. McConnell and Putnam operated the mine as early as 1854, and erected a twelve stamp mill. The last work on the property was done by the Cherokee Mining Company, the present owners, who suspended operations in 1901. This company did considerable work on both the east and west sides of the ridge. On the west side two shafts were sunk about one hundred and fifty feet apart on a quartz vein and connected by a drift at the hundred foot level. The more northwesterly of the two shafts is a hundred feet deep perpendicularly and a hundred and seventy-five on an incline. This shaft was drifted from in a north-easterly direction for about two hundred feet. The southwest shaft is approximately one hundred feet deep. Parties who have worked in the mine report that the vein averages about four feet in width, and carries large amounts of pyrite in the central portion. A fine grained hornblende schist was observed on the dump of one of the shafts above mentioned. Some mica schist was also noticed on the dumps and the latter rock is exposed in a considerable section in a large cut on the east side of the ridge. The association, therefore, at this point of the two classes of rocks referred to in Chapter IV is to be noted.

On the east side of the ridge a large excavation has been made by hydraulic work and a tunnel driven into the ridge for a considerable distance. It is reported that this tunnel cuts several veins on which drifts were run. A number of shafts are also to be seen on the east side of the ridge. A large mill is located on the property equipped with a ten stamp battery, two high speed rolls and a couple of Wilfley concentrating tables. An elaborately constructed cyanide plant is connected with the mill house. No work has been done at this mine for over six years. The mill house is in need of repairs and the shafts are inaccessible.

The results of four ore samples taken from the east side of the ridge and published in Bulletin No. 4-A, of the Geological Survey of Georgia, are given below. They were probably taken from the vein on which the two shafts previously described are now located.

No. 1.   Ore sample, Cherokee Mine,

.10 oz. ($ 2.00) of gold per ton.

No. 2.   Ore sample, Cherokee Mine,

.125 oz. ($ 2.50) of gold per ton.

No. 3.   Ore sample, Cherokee Mine,

.425 oz. ($ 8.50) of gold per ton.

No. 4.   Ore sample, Cherokee Mine,

.550 oz. ($11.00) of gold per ton.

General A. J. Warner, of Gainesville, Ga., one of the present owners, states that a car load of ore, shipped to Omaha for treatment, yielded over $200.00 per ton net.

SIXES MINE.—The Sixes mine, on lots 150, 212, 221 and 284, 15th district, first attracted attention by reason of the discovery of some very rich placer deposits along Sixes Creek. This deposit was among the earliest exploited placers of the Dahlonega belt, having been mined prior to the removal of the Cherokee Indians from the county. It is reported that many thousand pennyweights of gold have been obtained from an area of a few acres. In recent years Mr. T. C. Crenshaw, of Holly Springs, Ga., who was owner of the mine at the time of visit, has done some prospecting on veins, but at the time the property was examined the mine was closed and the shafts could not be entered. Two shafts have been sunk. One, not far from the mill house, is a two-compartment shaft sunk eighty feet on an incline of about 45°. A drift of approximately thirty feet has been driven from the bottom. About three hundred yards southwest of this shaft, another shaft has been sunk to a depth of one hundred and fifty feet and a little drifting done. This shaft is located at the contact of two rocks that, from field examination, appeared to be granite and hornblende schist. Large crystals of garnet were noticed in mica schist on the dump. This garnitiferous schist is said to show in the underground works as a narrow zone along the contact of the two rocks. As the shafts could not be entered, nothing can be stated concerning the veins. The mine is equipped with a ten stamp mill of 850 pound stamps, two Wilfley concentrating tables, a seventy-five horse-power engine and an air compressor of thirty drill capacity. The mill house and machinery are in excellent condition.

McCANDLESS PROPERTY.—Some prospecting on gold-bearing quartz veins was done on lot 61, 15th district, a number of years ago. From a quartz vein, showing a width of about a foot in some shallow excavations, a sample was taken for assay by State Geologist McCallie at the time Bulletin No. 4-A, of the Geological Survey of Georgia, was published. This sample yielded on assay $2.40 per ton.

CASE PROPERTY.—This property, lots 10 and 11, 15th district, has had considerable work done on it at different times. Shafts have been sunk and machinery installed. When visited, it was not in very good condition for examination, and nothing definite can be stated concerning the deposits.

LA BELLE MINE.—This mine is situated on lots 157, 204 and 205, 15th district. Some work on gold-bearing veins has been done in the past on all three of the above lots. On lot 205 a shaft was sunk to a depth of about eighty feet and some drifting was done. A small stamp mill was located here at one time, but the machinery has been removed. An open cut was made for a distance of about seventy-five feet on lot 175 and a shaft was also sunk and the ore taken out was milled at the mill on lot 205. Some prospect work was also done on lot 204, formerly known as the Casteel property. A sample of ore for assay taken from lot 157, when Bulletin 4-A, of the Geological Survey of Georgia, was published, yielded $2.44 per ton. One from lot 204 yielded $2.60 per ton.

PUTNAM MINE.—The Putnam mine is located on Blanket's Creek, lot 350, 15th district. Old placer workings covering about five acres at this locality are said to have been unusually productive. A pocket, or chute, of ore was found near the edge of the placer on a hillside which, according to reports, was wonderfully rich.

HAYNES PROPERTY.—Some mining was done on an auriferous quartz vein on this property, lots 348 and 349, 15th district, about six years ago. A two-compartment shaft was sunk to a depth of a hundred feet and a drift driven for sixty feet. The shaft seemed to be well timbered and is equipped with a boiler and hoisting engine. It was regretted that the underground works could not be entered when the property was inspected. The vein as exposed is said to average from fifteen inches to two feet in thickness. A sample for assay taken from an ore pile of a number of tons at the mouth of the shaft yielded $28.94 per ton. The mineral interest of the property is reported as being owned by Mr. J. W. Zachary, of Kentucky.

FARRAR MINE.—This mine is situated on lot 301, 15th district, and adjoins the Haynes property. It was formerly known as the Kolb property. A limited amount of prospecting on a quartz vein was done here many years ago. About 1907 an open cut, something like fifty yards in length, was made on a vein approximately four feet thick composed of quartz with interlaminated mica schist. A small test stamp mill was erected and from the ore milled it is reported that twelve or fifteen hundred dollars was secured.

CLARKSON MINE.—Some mining operations were carried on a number of years ago on lot 225, 15th district. Several tunnels were driven and a shaft was sunk to a depth of about eighty feet. Three samples for assay were taken from the vein at an exposure where it showed a width of about two feet, when Bulletin 4-A, of the Geological Survey of Georgia, was published. The results of these assays as given in that publication are as follows:

No. 1. Ore sample, Clarkson Mine,
.12 oz. ($2.40) of gold per ton.

No. 2. Ore sample, Clarkson Mine,
.125 oz. ($2.50) of gold per ton.

No. 3. Ore sample, Clarkson Mine,
.125 oz. ($2.50) of gold per ton.

MACOU PROSPECT.—Some prospect work consisting of shafts and open cuts has been done at two locailties on lot 158, 15th district. A stamp mill was in operation on the property for a short time in the eighties. Little can be learned at present as to the character of the deposits, but the ore bodies are probably in the nature of gold-bearing zones made up of thin layers or stringers of quartz with interlaminated schist. Pannings were made from ore material in the old works at both localities. A small amount of rather fine gold was obtained at each point.

WHORLEY MINE.—This mine is located on lots 459 and 460, 15th district, near the southwest edge of the Dahlonega belt. It is controlled by the Cherokee Mining and Milling Company, and at the time the property was visited, some prospect work was being carried on by Mr. Ed Billings, of Holly Springs, Ga.

In addition to old placer workings, considerable work was done in the nineties on an auriferous vein. A shaft was sunk on an incline of about 45° to a depth of one hundred and sixty-five feet and a drift run southwest for two hundred and twenty-five feet at a fifty-four foot level. A drift was also driven for about thirty-five feet at a one hundred and twenty foot level. As the underground works were not accessible when the property was inspected little could be learned as to the extent and character of the vein. The material of an ore pile of a number of tons at the mouth of the shaft appeared to be vein quartz and country rock interlaminated, or to consist of rock that had been impregnated with quartz and auriferous pyrite. Mr. Billings reports that the vein at the bottom of the shaft is something like six feet in width. A sample was taken for assay from the ore pile above mentioned from which about a ton of the richest ore had been culled. This sample yielded on assay $8.68 of gold per ton. A sample taken from about a ton of the richer cullings yielded $64.49 of gold per ton. There has recently been installed at the mine a small test stamp mill.

In addition to the principal mine, Mr. Billings has recently sunk several shafts prospecting on other veins on lot 406. Some gold was panned from a small vein exposed in one of these shafts, but until exploitation on these veins has been carried further, little can be said concerning them. Some tests were also in progress on placer deposits along a small stream flowing through the property. Efforts were made to examine this placer, but the gravels were difficult of access. Mr. Billings stated that two test pits that had been sunk yielded respectively sixty cents and fifty cents per cubic yard.

Owl Hollow, or Davis, Mine.—This mine is situated on lot 22, 15th district, about two miles south of the town of Canton. No work has been done at the locality for a number of years. As the property could be examined to better advantage when described by State Geologist McCallie in Bulletin No. 4-A, of the Geological Survey of Georgia, published in 1896, his remarks on the veins are here quoted in full:

"Three parallel gold-bearing quartz veins occur here, out-cropping about four hundred feet apart. The one furthest to the west has four shafts sunk on it, varying in depth from thirty to fifty feet. At the bottom of two of these, drifts have been extended along the vein for fifty feet or more, and the ore has been removed. The thickness of the ore body is quite variable in the different shafts, but at no place does it exceed two feet. The vein, consisting mainly of quartz, some-what laminated and frequently showing free gold, dips with the country rock at a high angle to the southeast.

"The east vein has one shaft eighty feet deep, sunk during the summer of 1893. Its upper portion resembles very closely the vein lying further to the west, but near the bottom, where it is said to attain a thickness of four feet, it contains much pyrite. A test of the ore, made on a Crawford mill erected near the shaft, was unsatisfactory. Whether this result was due to the low grade of the ore, or to the imperfect working of the mill we are uncertain."

RUDICIL MINE.—Placer mining was carried on along a small branch flowing into Mill Creek on lot 10, 2nd district, many years ago. Some prospecting for veins has also been done at different times on this lot and on lots 11 and 14. The locality is on the extreme southwest edge of the Dah-lonega belt. On the slope of the hill near the branch a limited amount of hydraulic work was done a few years ago. Water was brought by a ditch from a creek about three-fourths of a mile away and a reservoir constructed a short distance from the mine. The overburden, where a considerable excavation had been made, was very heavy and no gravels were noticed, but it was reported that a stratum of gold-bearing gravel occurred beneath the clay. It could not be ascertained what gold had been secured.

LATHAM MINE.—This property, lot 805, 3d district, is a short distance west of Orange postoffice. Some work was done here as early as 1852 and at different times since mining has been carried on by several parties. Three veins are said to occur here within a few yards of each other, and several shafts have been sunk and considerable drifting done. In the nineties a stamp mill was in operation, but as no work has been done for several years the veins could not be inspected when the property was examined. The results of two assays of samples from separate veins published in Bulletin No. 4-A, of the Geological Survey of Georgia, show values of $1.32 and $3.60 per ton respectively.

OTHER PROPERTIES IN THE VICINITY OF ORANGE POST OFFICE. —Prospect work on veins has been done on the property of Mr. Frank Burt a short distance Northeast of the Latham mine. The Sandow mine on lot 741 consists of old tunnels, shafts and open cuts on veins occurring in a ridge situated between Fowler and Smithwick creeks. Prospect work has also been done on lot 701. On the Richards property, lying on the opposite side of Fowler Creek from the Sandow mine, considerable prospecting was done at one time.

CREIGHTON, OR FRANKLIN, MINE.—This mine is situated in the extreme eastern part of Cherokee county on the Etowah River, and is about six miles distant from Ball Ground, a station on the Marietta and Knoxville Division of the Louisville & Nashville Railroad. Exploited veins at this locality traverse a number of land lots in the 3rd district, but the working shafts are distributed in a southwest-northeast course through lots 398, 465, 466 and 473. The surrounding country has a gently rolling topography, and the farming lands are productive and well timbered and watered. The external conditions here illustrate well the statement made in Chapter I of the advantageous mining location of the majority of the Georgia deposits. (Plate I, Frontispiece).

Considerable work was done at the Creighton mine before the Civil war by members of the Franklin family, the original owners of the property, and the mine was for many years known only as the Franklin mine. This early mining was confined chiefly to surface work. Oxidized ore was taken out at different points by open cuts and crushed at a small stamp mill located on the site of the present mill. Gold was also obtained from residual material in the vicinity of the ore bodies by sluice washing and with rockers. For several years after the Civil war the mine was owned by different companies and finally came into possession of the Creighton Mining Company who operated it more or less continuously for a number of years. The late Mr. Aaron French, of Pittsburg, widely known by reason of his many business connections, was one of the principal owners. The Creighton Mining Company first attempted to treat the sulphide ores by the cyanide method, but later a chlorination plant, still in use, was installed under the supervision of Mr. Adolph Thies. The mine is now owned by the Creighton Gold Mining Company, George C. Wallace, president, and Barry Searle, mining engineer and general manager. The Franklin Gold, Pyrite and Power Company, who own about six thousand acres along the Etowah River contiguous to the Creighton Gold Mining Company's property, transport and treat the ore of the Creighton mine at a reduction plant on the river a quarter of a mile to the north.

Several auriferous quartz veins having a parallel northeast-southwest trend occur near each other on the lots previously mentioned. The more important mining operations have been confined to one of these, known as the Franklin vein, it having been mined pretty continuously for three-fourths of a mile or more along its strike. This vein does not show a uniform width, but consists of successive ore chutes connected by quartz stringers. The average length of the chutes is something like seventy-five feet, and the thickness varies considerably at different points. The average thickness does not probably exceed three feet. The vein conforms in strike and dip to the country rock which has a strike of about N. 58° E. and dips toward the southeast. The ore chutes have a northeastward pitch. Mr. Searle, from his observations in the mine, finds the pitch of the chutes diagonal across the dip and general strike of the vein at an approximate incline of 30°.

The vein consists of milky quartz with varying amounts of pyrite, and at places contains considerable quantities of more or less mineralized wall rock as interlaminated bands or layers. Gold in the quartz dissociated from pyrite is not infrequently noticed. The predominant sulphide is iron pyrite, though a little chalcopyrite is found. The pyrite in some portions of the vein is in large amounts and occurs in distinct bands of varying thickness parallel with the walls of the vein and alternating with layers of quartz. Plate II, Fig. 2, gives a view of some specimens of this striking looking banded ore.

The veins at the Creighton mine occur near the southwest edge of a body of rock, composed, where exposed, of hornblende schist and mica schist or quartz-feldspar-mica schist. This formation can be traced for a quarter of a mile, or more, across its strike to the northeast, the soil from its residual material being of a deeper, more reddish, color than that derived from other rocks of the region. Southwest from the mine quartz-feldspar mica schist outcrops at one or two localities and is succeeded by a body of gneiss that yields on weathering a characteristic light gray soil. The hornblende schist and mica schist, where exposed, are usually so intimately mixed that they could not be mapped as separate series of rocks. The close similarity of the hornblende rock, however, with other hornblende rocks occurring as distinct bodies on both the southwest and northeast extension of the gold belt, leaves little doubt that the association, found very commonly along the Dahlonega belt, of a basic with a more acidic type of rock, is represented here. The absence of distinct differentiation at this locality may be due to an intimate mixture of original rock magmas; or the limited exposures may be along ancient contacts presenting unusual features.

An interesting feature in connection with the rocks at the Creighton mine is the occurrence of an olive diabase dike cutting the Franklin vein in shaft No. $3\frac{1}{2}$ a little below the two hundred-foot level. This dike is over a foot in thickness and is the best example that was found on the belt of a series of younger intrusive rocks cutting the more ancient quartz veins. No displacement of the vein seems to have been caused by its intrusion.

Five working shafts, designated by numbers, are located at intervals along the strike of the Franklin vein for nearly three-fourths of a mile. Shafts Nos. 1 and 2 are near together and only a short distance from the south bank of the Etowah River; shaft No. 3 is about fourteen hunderd feet southwest of No. 2; shaft No. $3\frac{1}{2}$ is about seven hundred feet southwest of No. 3; and shaft No. 4 is about sixteen hundred feet southwest of No. $3\frac{1}{2}$. Southwest of these shafts the outcrop of a quartz vein that is probably the continuation of the Franklin vein has been traced for some distance. At a point where the float ore is unusually promising looking, a shaft was sunk by the management preceding the present one, with the intention of exploiting the vein. Unfortunately, for business reasons, the work was ordered stopped just before the depth was attained at which the vein was expected to appear.

The major developments of the working shafts are along the incline of the ore chutes. From shafts Nos. 1 and 2 nine or ten drifts have been run and four or five ore chutes mined to a depth of about five hundred feet vertical. An approximate depth of nine hundred feet has been reached by these shafts measured along the incline of the ore chutes. Shaft No. 3 has an inclined depth of about three hundred feet. Shaft No. 3½ has an approximate depth of five hundred feet along the incline of the ore chute on which it is located. Shaft No. 4 has an approximate inclined depth of nine hundred feet.

At the time the property was visited the only ore bodies in the mine accessible for sampling were those reached from shafts Nos. 3½ and 4. As samples had been taken systematically at these two localities in 1907 by State Geologist McCallie and the assay results published in a special report submitted to the Central Bank & Trust Corporation, of Atlanta, Ga., it was deemed unnecessary to duplicate those samples. The assay results as published are given below.

SHAFT No. 3½

Sample No. 1, bottom of shaft, vein 30 inches, value .76 oz. ($15.71) gold per ton.

Sample No. 2, 56 feet from bottom, vein 28 inches, value only a trace.

Sample No. 3, 97 feet from bottom, west end of drift, vein 16 inches, value .35 oz. ($7.23) gold per ton.

Sample No. 4, 97 feet from bottom, east end of drift, vein 18 inches, value .66 oz. ($13.64) gold per ton.

Sample No. 5, 97 feet from bottom, center of drift, vein 41 inches, value .25 oz. ($5.17) gold per ton.

Sample No. 6, 147 feet from bottom, west wall of shaft, vein 30 inches, value 1.20 oz. ($24.80) gold per ton.

Sample No. 7, 215 feet from bottom, east wall of shaft, vein 18 inches, value .74 oz. ($15.30) gold per ton.

Average value of above several assays, $11.70 gold per ton.

SHAFT No. 4

Sample No. 1, east end of drift, bottom of shaft, vein 9 inches, value 1.09 oz. ($22.53) gold per ton.

Sample No. 2, center 60-foot drift, bottom of shaft, vein 34 inches, value 1.27 oz. ($26.25) gold per ton.

Sample No. 3, west of drift, bottom of shaft, vein 12 inches value .40 oz. ($8.27) gold per ton.

Sample No. 4, top of slope 37½ feet above 758-foot level, vein 36 inches, value .74 oz. ($15.30) gold per ton.

Sample No. 5, east end of drift, 758-foot level, vein 12 inches, value .50 oz. ($10.34) gold per ton.

Sample No. 6, west end of drift, 758-foot level, vein 12 inches, value only a trace of gold.

Sample No. 7, east side of shaft, 40 feet above 677-foot level, vein 48 inches, value 1.49 oz. ($30.80) gold per ton.

Average value of above several assays $16.21 gold per ton.

From the northeast end of the works on the Franklin vein at a depth of about two hundred feet from the surface a tunnel was driven some years back to a parallel vein approximately one hundred and seventy feet distant, known as the McDonald vein. From this tunnel a drift was run on the strike of the latter vein for about one hundred and fifty feet. No extensive operations, however, have been conducted on this vein, and as the works above referred to were inaccessible at the time the mine was visited, nothing can be stated as to its probable value.

In addition to the Franklin and McDonald veins, some other veins have been located on the property, but they have as yet been prospected to only a limited extent.

The probable extension of the Franklin vein for a considerable distance southwest of shaft No. 4, as judged by rather strong outcroppings, would seem to warrant more extensive prospecting southwestward than has heretofore been undertaken.

The ores from the Creighton mine are treated at the Franklin Gold, Pyrite and Power Company's plant, which is situated on the Etowah River about one-fourth of a mile to the north. Mr. Barry Searle is general amnager here as well as at the mine. A power plant for operating the drills, electric lighting, etc., is on the east bank of the river. The plant for treating the ores, consisting of a mill and chlorination plant, is located on the west bank. Water power, obtained in abundance by a dam across the river, is utilized in operating both plants.

The mill is equipped with a stamp mill of the ordinary type. Frue vanners are used principally in concentrating the sulphides. It is estimated that about one-half of the values in the ore are secured by amalgamation on the plates.

A three-story chlorination plant is located convenient to the mill house. The concentrates, after thorough roasting in a furnace on the first floor, are elevated to the third story and conveyed by a car to a number of iron chlorinating barrels lined with lead. Each barrel, after receiving the proper amount of roasted concentrates together with a chlorinating charge, is revolved slowly on its axis for a sufficient length of time to allow the gold to be converted into a chloride. The pulp is then removed to filter presses and the chloride of gold leached out and conveyed to storage tanks. It is stated that, on account of some of the gold having been in too coarse particles to go into solution, a large amount of material at this plant, accumulated from former operations, still contains several dollars of gold to the ton.

At the time the property was visited, a more substantial dam than the old one was in course of construction across the river immediately above the site of the mill house. New water wheels were also being installed at the mill, and other improvements were in progress under the able supervision of Mr. Searle.

Cox Property.—This property, located about a mile and a half west of the Creighton mine, was prospected for gold a number of years ago. Two veins of quartz interlaminated with layers of schist were exposed in a short tunnel. Several shallow pits and open cuts were also made at the same time. A small test mill was operated for a time on the property.

Other Properties in Cherokee County on the Dahlonega Belt.—In addition to the properties already described, prospect work has been done on lots 208 and 971 in the 15th district, on lots 721, 723, 760, 829, 826 and 959 in the 21st district, and on lot 848 in the 20th district. Some prospecting has also been done on a quartz vein a few miles northeast of the Creighton mine at the Hendrix place. Old placer works are to be found along Downing's Creek on lots 63, 64 and 81 in the 15th district.

FORSYTH COUNTY

The Dahlonega belt embraces an area of a few miles in extent in the extreme northeast corner of Forsyth county. Only one property in this area has been developed sufficiently to merit description.

CHARLES MINE.—This mine is on lot 77, 3rd district, 1st section. Quite an amount of work has been done here in the past, and a stamp mill is still located on the property. Two parallel auriferous quartz veins are said to occur near each other and, in addition to the underground works near the site of the mill house, prospect work has been done along their strike for several hundred yards by means of pits and open cuts.

No work has been in progress for a number of years and the veins could not be examined very satisfactorily. In one or two excavations a quartz vein from two to three feet in thickness was noticed. The occurrence at this locality of arsenopyrite in the vein quartz, noted by several observers, is to be mentioned.

DAWSON COUNTY

The Dahlonega belt traverses Dawson county diagonally, passing within about a mile of Dawsonville, the county seat. Its cross dimension in the southwest corner of the county probably does not exceed two miles, but before Lumpkin county is reached to the northeast a considerable widening occurs.

Less extensive gold mining operations have been prosecuted in Dawson county than in either Cherokee county on the southwest or Lumpkin county on the northeast. This lack of activity in mining is not readily explained as the same class of deposits are found here as in the other two counties.

PROPERTIES NEAR BARRETTSVILLE.—The Barrett Gold Mining Company did some work for gold on lot 1,107, 4th district, about eight years ago. Mr. C. A. Vandiviere, of Dawsonville, has also done some prospect work in the same neighborhood on lot 1,104, 4th district.

KIN MORI, OR HARRIS BRANCH, MINE.—This mine is situated on Harris Branch about four miles south of Dawsonville. The property includes lots 858, 859, 862, 908, 909, 910, 911, 926, 927, 928, 929, 976, 977, 978, and 979 in the 4th district.

The most extensive gold mining operations that have been conducted at any point in the county were carried on at this locality during the eighties by a company known as the Kin Mori Gold Mining Company. The first work consisted of the mining of placer deposits along Bean Branch, and in the Etowah River bottoms above Hugh's Shoals. Water for these operations was brought by a ditch, thirty miles or more in length, from Nimble Will Creek in Lumpkin county. Later, a thirty stamp mill was erected on Harris Branch and the saprolite deposits of the so-called Quarles belt, situated some distance above the site of the mill, were extensively mined by hydraulic methods. Several large cuts were made in the decomposed material of this belt which consists of schists carrying many small stringers and lenses of quartz. Dr. Becker, in the paper that has been a number of times previously referred to, notes at this locality as many as twenty tiny quartz veins in a mass an inch in thickness[1]. The material displaced by hydraulic giants was carried through a long line of sluice boxes to the mill where the coarser portions were retained in a bin and fed to the batteries. As is usually the case in this method of mining saprolite deposits, a large percentage of the total values secured were saved in the sluice boxes.

The mill house and the greater portion of its machinery, both in a very dilapidated condition, are still to be seen on Harris Branch.

A rather large quartz vein occurs on this property known as the "Big Sulphuret Vein." It has been prospected along

---

1. Becker, G. F., Gold Fields of the Southern Appalachians: Sixteenth Annual Report, U. S. Geol. Survey, pt. 3, p. 296.

its strike for some distance from a point near the mill. It is probably of low grade, though a sample taken from a tunnel on it, the assay of which was published in Bulletin No. 4-A, of the Geological Survey of Georgia, showed a value of $23.40 per ton. Mr. Gazzam Gano, of Cincinnati, is at present in charge of this property.

OTHER PROPERTIES NEAR THE KIN MORI MINE.—The McGuire property, consisting of lots 912 and 925 in the 4th district, 18th section, joins the Kin Mori, or Harris Branch, property on the east. A small vein was discovered here a number of years ago which is said to have yielded several thousand pennyweights of gold with a very limited amount of work. Another vein, known as the Whippoorwill vein, has also been located near by.

Some prospect work has been done in the past on lot 1,041 of the Looper property in the same district.

MAGIC MINE.—This mine is northeast of the Kin Mori Mine, on the side of a ridge known as Spike Hill. The property embraces lots 369, 370, 376, 424, 425, 432, 483 and 490 in the 13th district (north half).

Considerable work was done at this locality a number of years ago by the Kin Mori Gold Mining Company. The operations appear to have consisted principally of hydraulic work in saprolite material. What gold was secured here was not ascertained. Some buildings, erected at the time the mine was in operation, are still on the property.

ELLSWORTH, OR FRACTION, MINE.—This mine is about two miles east of Dawsonville on fractional lot 54, 14th district. A very limited amount of work was done at this locality a number of years ago by a company known as the Ellsworth Mining Company. Little could be seen in the old excavations of the material mined, but the deposit is said to consist of a zone about two feet in thickness composed of schist and stringers of quartz. Where this zone was worked, near a small branch, it dips under a hill, and in addition to a small excavation in the hillside, an incline shaft was sunk on the dip of the zone for about fifty feet. A limited amount of drifting was also done from this shaft. The company erected a small stamp mill near the shaft; the building is still standing, but the mill has been removed.

OTHER PROPERTIES IN THE VICINITY OF ELLSWORTH AND MAGIC MINES.—Lots 60, 61, 62, 63 and 547, 13th district, and lots 53, 646, 647, 648 and 714 in the 4th district, are included in a property known as the Amicalola Mine upon which a small amount of work has been done. The same may be said of the Missing Link Mine on lots 373, 427, 430, 431, 483, 484, 489 and 490, 13th district (north half).

LOTS 366 AND 376.—In the northeast corner of lot 376, 15th district (north half) a quartz vein averaging several feet in thickness has been stripped for a distance of about fifty feet. The quartz appeared rather promising looking, but a sample taken along the exposure and assayed in the laboratory of the Survey yielded no gold.

On a ridge on lot 366 in the 13th district (north half) a large quartz vein outcrops rather strongly, and a few shallow test pits have been sunk at one or two points along its strike. This vein is supposed to be the continuation of a large vein exposed on the Church lot and the Shelton property. A small excavation has recently been made on a vein on lot 366 which may possibly be the large vein just mentioned, but the appearance of the ore would indicate a different vein. From this cut it is said that about eighty pennyweights of gold was secured. The deposit at this locality is known as the Cooper prospect.

CHURCH LOT.—This lot is about four miles northeast of Dawsonville in the 13th district (north half). A large quartz vein five feet or more in thickness has been exposed in a small pit a short distance back of a church building. A sample for assay was taken from a section across this vein, but the assay showed no gold.

SHELTON PROPERTY.—On lot 241 of Mr. J. F. Shelton's property, not far from the last described locality, a large quartz vein is exposed in the channel of a branch. A little work on it has also increased the extent of the natural exposure. This vein, like the one on the Church lot, is of large dimensions, and it is quite probable that it is a continuation of that vein. Some portions of the vein where it is exposed show large amounts of iron pyrite. A sample for assay was taken from a section across the vein, but it yielded no gold. The results of an assay from a sample from the same exposure of this vein, published in Bulletin No. 4-A, of the Geological Survey of Georgia, shows a value of $1.00 per ton.

PALMOUR PROPERTY.—Some work was done several years ago on an auriferous vein on lot 361, 13th district (north half) which yielded some quite rich ore. An open cut was made on the vein and a small stamp mill was erected with which a limited amount of the ore was milled. As no work has been done at the locality for some years and little could be seen of the deposit, a description of the veins published in Bulletin No. 4-A, of the Geological Survey of Georgia, in 1896, is quoted below.

"In the cut, and well separated from each other, are a series of quartz stringers averaging from one to six inches in width. On the right side of the cut one of these stringers has been discovered to be exceedingly rich in gold. This lies close to the right hand wall, and has varied from a thin ribbon

to a vein one foot in width. It is this vein which has furnished
the major portion of the gold found by all workers here, and
it is this vein for which the excavation was made.

"The vein is quite typical in appearance and may be
readily distinguished from the others as far as it has been
operated. It is made up of a dark, finely granular quartz
arranged in thin parallel bands of about one quarter of an
inch in thickness. This laminated structure of the vein
material causes it, when picked out by the miners, to break
into rectangular pieces; and on this account it is referred to
as "The Palmour Brickbat Vein." Portions of the vein run
extremely high in free gold, the gold occurring not only in the
quartz matrix, as I found by powdering and pounding a considerable quantity, but especially along the lines of lamination.
In the work conducted under the Survey management, a large
number of pieces were split open, and almost invariably flaky
particles of gold were apparent to the naked eye. Several
pieces have been found incrusted with such flakes. The rotten
schhist walling was also pounded freely, and from each small
panful of earth the color obtained was surprising."

Considerable placer working has been done near by on
Proctor's Creek. These placer deposits are reported to have
yielded good returns. The property is owned by Messrs. D.
M. McKee and N. D. Black.

LOTS 264 AND 297.—Some hydraulic work on a placer
deposit at the mouth of Long Branch on lot 297, 13th district
(north half), had recently been in progress when the region
was visited. This work was done by the Etowah Gold Mining
and Ditch Company, but it could not be ascertained how
successful their operations were. The same company had
also been recently carrying on some hydraulic mining on a
deposit in a dry hollow or gulch and on the adjacent hill slopes,
on lot 264 in the same district. Some work was also done on

lot 235 in the 13th district. Water for these operations was brought from Lumpkin county by a ditch of considerable length.

LUMPKIN COUNTY

The Dahlonega belt passes diagonally through the southern half of Lumpkin county and has an average width of about six miles. Dahlonega, the county seat of Lumpkin county and the town from which the belt takes its name, is situated in the midst of extensively exploited properties. More capital has been invested and operations of a more extended character carried on in mining for gold in Lumpkin than in any other county in the State.

The working of saprolite deposits and gold-bearing zones by hydraulic methods, as described on page 31, to a depth at which the country rock is too resistant to yield to the cutting power of water has been a favorite form of mining in this region. The particular variety of mining operations just mentioned, so much in vogue in this county, became known years ago throughout the Appalachian gold fields as the "Dahlonega Method." Hundreds of thousands of cubic yards of decomposed rock have been removed at a number of localities in the vicinity of Dahlonega in the course of these hydraulic operations. Plate IV, Fig. 1, gives a view of a portion of one of the huge cuts to be seen in this section.

The large cuts that have been made at a number of points in the vicinity of Dahlonega afford unusual opportunities for studying the geology of the deposits. Unfortunately, the structural relations of the rocks at this locality are very complicated and difficult to interpret. The occurrence of the basic hornblende series of rocks, previously mentioned, in connection with more acidic rocks, is very noticeable in the Dahlonega district. The presence also at Dahlonega, within the boundaries of the gold belt, of a considerable body of granite and of several sheared granite dikes probably connected with it, is a matter of interest. This granite body, while more or less sheared, is not as schistose as the majority of the rocks of the region and is probably of more recent age than the others.

ETOWAH MINE.—This mine is located on lots 117, 118, 119, 120, 141 and 178, 15th district. The property is along the Etowah River and the Dawson-Lumpkin county line. Considerable work has been done in the past on placer deposits along the Etowah River, and excavations of some magnitude have also been made in the saprolites with hydraulic giants. No work was in progress when the property was visited and nothing definite can be stated concerning the deposits.

PARKER LOT.—This property adjoins the Etowah property. Some excavations were made on the summit of a ridge at this locality by hydraulic work many years ago. Recently a tunnel was driven into the hill from near its base for the purpose of cutting the veins. As no work was going on when the property was inspected nothing could be learned as to what success had attended these operations.

GOLD HILL MINE.—This property is only a short distance from the two properties previously described. Considerable work was done on a hill at this locality years ago. Hydraulic work was carried on at several points, and open cuts were made and some tunnels driven into the hill. Recently the Etowah Gold Mining and Ditch Company made an excavation of considerable magnitude by hydraulic work on one side of the hill. What gold was obtained in these later operations could not be ascertained.

JOSEPHINE MINE.—This mine is located on lots 526, 595 and 1,215, 12th district, and lots 17, 18, 48, 49 and 82, 13th

PRELIMINARY MAP OF DAHLONEGA DISTRICT, GEORGIA
BY
ARTHUR KEITH.

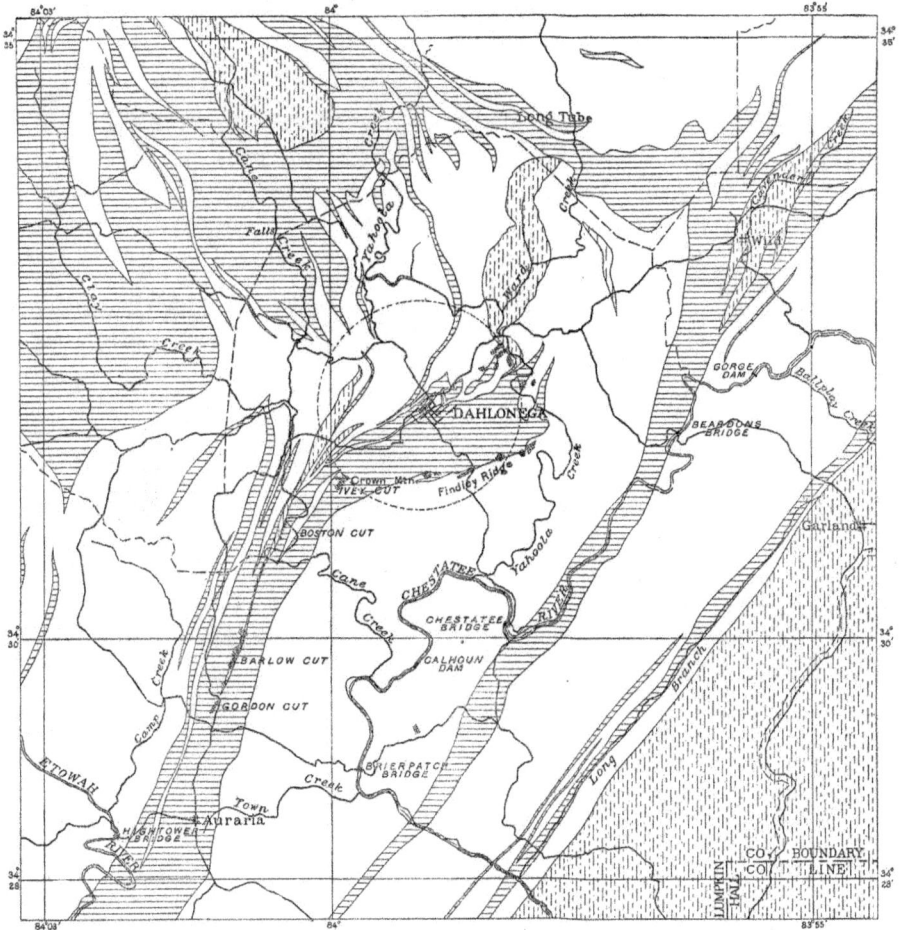

Topographic base by
U.S. Geological Survey

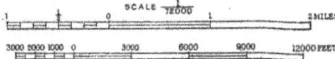

SCALE

3000 2000 1000 0   3000   6000   9000   12000 FEET

1909.

Carolina gneiss	Roan gneiss	Granite	Open Cut
(mica-gneiss, mica-garnet-ottrelite-and black slaty schists)	(hornblende-gneiss and schist, diorite, and metagabbro)	(massive and schistose granite)	

district (north half). Placer deposits here along the Etowah River and McCluskey Branch have been extensively worked in former years. The placer deposits along McClusky Branch are reported as having been unusually productive. In addition to placer mining considerable work was done some years ago on a hillside. These operations consisted of excavations in the saprolites with some tunneling and sinking of shafts. A mill house with part of a stamping outfit is still on the property, but no mining operations have been in progress for a number of years and nothing definite can be stated concerning the character of the deposits.

BATTLE BRANCH MINE.—This mine is on lots 457 and 524, 12th district, on the west side of the Etowah River. The mine is located on a small branch known as Battle Branch, and has been worked at intervals by different parties since 1831. When the property was visited, work, which had been going on for several years under the direction of the present management, had lately been suspended. The mine is reported as being controlled by the Battle Branch Gold Mining Company, A. E. Rodgers, of Boston, Mass., being president.

In addition to placer work along the branch some large cuts have been made in the saprolites and shafts sunk. The shafts could not be entered when the property was visited, but it is stated by reliable parties that some extremely rich pockets have been worked during the development of this mine and a great deal of gold secured.

BETZ MINE.—The Betz mine is located on lot 388, 12th district a short distance from the Etowah River. It is, as are also the two previously described properties, in the vicinity of Auraria. This mine was formerly known as the Wing Mine. It is now being operated by the Piedmont Mining and Milling Company, Mr. Charles P. Tasker, of Philadelphia, being general manager.

In the course of the different mining operations that have been carried on, a large excavation has been made in partially decomposed mica and quartzose mica schist, and several incline shafts have been sunk from the bottom of the excavation along the incline of auriferous zones or strata of schist containing stringers of quartz. Considerable drifting and stoping have also been carried on from the shafts. At the time the property was visited the old workings had been only partially cleaned out by the present management, and little could be seen of the ore material in the drifts. It is stated, however, that there are large bodies of auriferous schist exposed in the open cut that will pay to mill. This material which is partially decomposed, is broken up in an ore beater and carried by a flume line to the mill. The following assay results are quoted from Bulletin No. 4-A, of the Geological Survey of Georgia, published in 1896. The assays were made from samples taken at that time by the Survey from four auriferous belts or strata at this mine. These belts are described in that publication as follows: "Vein No. 1 is four feet thick; vein No. 2, eight feet thick; vein No. 3, twenty-eight feet thick; and vein No. 4, thirty-four feet thick." The assays as given are:

Vein No. 1_____ 0.105 oz. ($2.17) of gold per ton.
Vein No. 2_____ 0.150 oz. ($3.10) of gold per ton.
Vein No. 3_____ 0.070 oz. ($1.45) of gold per ton.
Vein No. 4_____ 0.080 oz. ($1.65) of gold per ton.

The milling plant of the Betz mine is situated a short distance below the mouth of the open cut, the ore, as previously stated, being conveyed to the mill by a flume line. At the time the property was visited the milling equipment consisted of two Huntington mills with double amalgamating plates, a stamp mill of ten stamps and two Wilfley concentrating tables.

FIG. 1.—MILLING PLANT, PARKS GOLD MINE, McDUFFIE COUNTY, GEORGIA

FIG. 2.—INTERIOR VIEW OF STAMP MILL SHOWING BATTERY AND AMALGAMATING PLATES, PARKS GOLD MINE, McDUFFIE COUNTY, GEORGIA

The portion of the ore adapted to the Huntington mills is conveyed from the flume line into a bin from which it is fed to these mills. The coarser and more resistant portions of the ore are treated in the stamp mill. The machinery is in excellent condition and is operated by electrical power brought from a power plant situated a short distance from the mine on the Etowah River.

It will be noted that the average of the four assays quoted above is, in round numbers, $2.00 per ton. The combined thickness of the four auriferous zones from which they were taken is seventy-four feet. These tests are entirely too meagre to draw satisfactory conclusions concerning the value of such a large body of material. Yet, taken in connection with the history of the mine and the statements made concerning it, they point encouragingly to the prospect of a large body of low grade ore. If the auriferous zones continue along their strike further than they have been exposed in the present cut, there will be a large amount of material that can be mined very cheaply, owing to surface decomposition.

Future developments at this mine will be looked forward to with interest, as extensive, and economically conducted work on large bodies of low grade ore offer better prospects for permanent success than any other class of mining operations afforded by the Georgia gold fields.

McINTOSH LOT.—The McIntosh lot, 386, 12th district, has had some prospect work done on it. It adjoins the Betz lot on the south and it is claimed that the auriferous belt at the Betz mine passes through this property.

LIBERTY BELL MINING COMPANY'S PROPERTY.—Since field work was completed in the preparation of this report, it has been learned that the Liberty Bell Mining Company is making rather extensive preparations for hydraulic mining at a locality about a mile from Auraria between the forks of the Gainesville and Dawsonville roads.

NORRELL MINE.—This mine is on lot 736, 12th district, about a mile southeast of Auraria. Some mining operations on veins and saprolite deposits were carried on here years ago. In the eighties, Mr. John Norrell, the original owner of the property, sunk some shafts and did a limited amount of work on the auriferous veins. About 1893 Messrs. Stewart and Woodward carried on some mining operations at the locality for a year and a half, making an open cut something like seventy feet in length. A gold-bearing zone about a hundred and fifty feet wide is said to traverse the lot.

HEDWIG MINE.—This mine, formerly known as the Chicago and Georgia Mine, is located about one-fourth of a mile north of Auraria and a little to the west of the public road between Auraria and Dahlonega.

In addition to early placer workings along several small branches, some large excavations were made in the saprolites eighteen or twenty years ago. A company, known as the Chicago and Georgia Company, made by hydraulic work a large excavation, designated as the Chicago and Georgia cut, in the side of a hill on lot 663, 12th district. Two auriferous zones, reported as being thirty and fifty feet thick, consisting of schists, carrying numerous quartz stringers, are said to have constituted the ore bodies. This company also erected a twenty stamp mill for the treatment of the ore. Later, under the ownership of Mr. Christian Wahl, another cut was made on the same lot to the east of the Chicago and Georgia cut known as the Hedwig cut. Other work was also done on the Chicago and Georgia cuts. In addition to the mining operations already described, some work has been done on lot 662, adjoining lot 663. The mine is owned by the Wahl estate.

WHIM HILL MINE.—This mine is situated on lot 670, 12th district, a short distance north of the Hedwig mine. Considerable work was done here on the crest of Whim Hill, immediately west of the road from Auraria to Dahlonega. Several shafts were sunk and some tunnels driven. The work was done many years ago and nothing can be stated concerning the character of the veins. In the early workings, a whim was used in sinking one of the shafts and from this circumstance the mine and the hill received their names. Mr. Elisha Castlebury worked the mine in the forties and his grandson, Mr. J. F. Castlebury, of Dahlonega, and other parties familiar with the mining history of the region, state that some ore chutes were encountered that were exceedingly rich in free gold. The lot on which the mine is located forms a part of the Wahl estate.

BRIAR PATCH AND CALHOUN MINES.—These two mines are now part of an extensive property controlled by The Dahlonega Gold Mining and Milling Company, Mr. B. L. Payne, of Lincoln, Nebraska, president, and Mr. P. P. Carmichael, local manager. What was originally known as the Calhoun mine is on the east side of the Chestatee River on lots 164 and 165 in the 11th district, and the Briar Patch mine, embraced a number of forty acre lots on the opposite side of the river. The lots now included in the property are: Nos. 164, 165, 169, 170 and part of 168 on the east side of the Chestatee River, and Nos. 1191, 1192, 1193, 1194, 740, 800, 801, 802, 803, 808, 809, 871, 872, 873, 878, 879 and 880, all smaller forty acre lots, on the west side of the river.

The Calhoun mine is rather widely known as being the locality where gold was first discovered in the Dahlonega region, and several very rich ore chutes have been worked here in the past down to water level. Some open cuts were made and shallow shafts sunk years ago on several small veins or chutes near together on lot 164. John C. Calhoun, the noted

South Carolina statesman, acquired possession of the property soon after the presence of gold had been discovered. During his ownership, the mine was worked under a lease for thirty days by two parties, W. G. Lawrence and Charles Sisson, and it is stated on reliable authority, that they secured in that short period $24,000 worth of gold. None of the old workings probably extended much below water level.

At the time the property was visited, Mr. Carmichael was cleaning out an old inclined shaft on what was supposed to be one of the richest of the ore chutes. A small quartz vein, averaging less than a foot in thickness, where exposed, and dipping at a low angle, could be followed for several yards in this shaft. At, or a little below, water level, some ore very rich in free gold was noticed. At one point on the vein free gold was quite abundant in a narrow band or layer of exceedingly rich quartz traversing the vein diagonally. In addition to the work just mentioned, Captain Fry, of Dahlonega, was driving a tunnel into the hillside to strike the Loggin chute, another ore chute on lot 164 that is reported as having yielded rich returns from former workings.

A short distance from the ore chutes above mentioned, an auriferous quartz vein occurs known as the Peachtree vein. This vein has been worked to a limited extent at a number of points along its strike, and its continuity for some distance has been pretty well established. Owing to the limited extent of exposures when the property was visited no samples for assay were taken from this vein.

In addition to the work that has been done on veins at the Calhoun mine, considerable work on placer deposits in the Chestatee River bottoms has been carried on. According to the best information obtainable, the placer mining here was quite profitable.

On the west side of the Chestatee River there is a large

body of alluvial land traversed by two or three small streams flowing into the Chestatee. Placer deposits occurring here have been known for many years as the Briar Patch Mine. Placer mining was carried on at this locality quite vigorously in the early mining days. It is frequently difficult at some of the placers that have been extensively worked to determine what areas have been mined and what are still virgin ground. It is claimed, however, that, owing to heavy overburden and slight drainage fall, a considerable portion of the river bottoms at this locality has never been worked. Mr. Carmichael is of the opinion that there are large areas here of alluvial deposits that have never been mined.

On lot 803, a short distance west of the river, a hill known as Gold Hill, has been the scene of some recent interesting prospect work. A number of small veins or stringers have been located on the top of this hill and exposed in shallow pits. Until some sluicing work is done, or larger, more continuous excavations made, nothing very definite can be learned concerning these deposits, but the number and location of the small veins that can be examined suggest the presence here of a gold-bearing zone of considerable dimensions. Pannings were made from several of these small veins or stringers. These stringers of quartz break up into small pieces when dug out of the saprolites and a portion of the decomposed walling was included in the panning tests. The fragments of quartz were saved in the first pannings and subsequently mortared and panned. In both tests the results were very gratifying, considerable amounts of fairly coarse gold being obtained.

A water ditch a number of miles long has just been completed for bringing water to the Dahlonega Gold Mining & Milling Company's property. This ditch is designated on maps of the region as the Briar Patch ditch.

A forty-stamp mill and a large hydraulic pump for sluicing

operations are located on the Chestatee River at the corners of lots 164 and 168. A dam across the river, developing eight hundred horse-power, gives water power for operating this plant. The company also owns a dredge boat on the Chestatee River, which was anchored, at the time the property was visited, near the Briar Patch bridge.

TURKEY HILL MINE.—This mine, situated on lot 163, 11th district, adjoins the Calhoun mine on the south. Near the crest of a high ridge an open cut was made for a distance of about two hundred feet a number of years ago. Within this cut some shafts have also been sunk. No work has been done at the locality for a number of years and the shafts could not be entered. The following description of veins occurring here is quoted from Bulletin No. 4-A, of the Geological Survey of Georgia, published in 1896, at which time the mine could be examined to much better advantage:

"This cut exposed five veins, which are about fifteen or twenty feet apart, varying from one to six inches in thickness. These veins are very rich and show considerable free gold in a porous, saccharoidal quartz very slightly stained throughout to a dirty cream color by iron sesqui-oxide."

There are at present in the collections of the State Museum in the State Capitol, some very handsome specimens of rich gold quartz from this mine that were donated by the owners a number of years ago. The property is controlled by Mr. F. S. Packard.

McAFEE-LYNN MINE.—This mine lies a short distance to the northwest of the Briar Patch mine. It was originally worked for placer deposits along Ralston Branch and was called the Rutherford mine after one of its former owners, the late Prof. William Rutherford, of the University of Georgia. Subsequently the property came into the possession of

Messrs. McAfee and Lynn and is now more generally known as the McAfee-Lynn mine. About 1904 some rather extensive mining operations were carried on at this locality. A large cut was made in the saprolites known as the McAfee-Lynn cut.

No work was going on when the region was visited and here, as at other large cuts in the Dahlonega region where no mining was in progress, an examination was very unsatisfactory. A stamp mill is located on the property.

BARLOW MINE.—The Barlow mine is about three miles southwest of the Dahlonega public square and between a quarter and a half of a mile to the east of the public road from Auraria to Dahlonega. Very extensive hydraulic mining operations were formerly carried on at this locality. Several, originally, separate properties, the Ralston mine, the Pigeon Roost Mine, and others, have now been consolidated into one mine, the different cuts having merged into each other or become closely connected. Lots 746 and 747 in the 12th district have been the center of some of the most important operations, but a number of different land lots are included in the property.

The principal mining here has consisted of hydraulic operations, and a very large amount of saprolite material has been removed in the course of the work. Plate IV, Fig. 1, gives a view of a portion of the immense Barlow cut, which is a half mile, or more, in length. Separated from this cut by a thin partition and branching off toward the northwest is another cut. Near the end of this, farthest from the Barlow cut, an old shaft has in recent years been cleaned out and deepened and some underground work done on a vein known as the Ogle vein. The cut to the northwest is a part of the old Pigeon Roost mine.

The first vein mining done at this locality was conducted by a company known as the Georgia Company, which, in 1866,

erected a forty-stamp mill. Pretty continuous mining was
carried on during the seventies and eighties. At one time a
sixty-stamp mill was in operation near one end of the
extensive system of cuts and a twenty-stamp mill was erected
at the opposite end. Water was brought by a branch-ditch
from the large Hand ditch near Dahlonega. A forty-stamp
mill now occupies the site of the old sixty-stamp one.

With the cuts in their present condition, little can be
ascertained concerning the veins. Judging, however, from
the character of the operations that were carried on and from
general reports the deposits are probably in the nature of
zones or belts of schists carrying auriferous quartz stringers.
The association of the hornblende series of rocks, several
times previously referred to, with an acidic schist representing, at this point, probably a highly sheared granite porphyry
is noticeable in the large cut.

About 1906 the Water Power and Mining Company of
Georgia cleaned out and deepened a shaft on the Ogle vein
near the mouth of the northwest cut. Some drifting was done
from this shaft, but the underground works were inaccessible
at the time the property was visited and nothing can be stated
concerning the character of the vein.

According to general report, the output of gold in the past
from the Barlow mine has been very large, but unfortunately
definite statistics are not obtainable. The property is owned
at present by the Water Power and Mining Company of
Georgia.

GORDON MINE.—This mine is about two miles southwest of
Dahlonega court house and is situated on lot 791, 12th district.
Work was done here as early as the forties, a wooden stamp
mill having been erected on Cane Creek, not far from where
a large cut, known as the Boston cut, is located. In addition
to the cut just mentioned some surface work has been done

on the hill to the north of the Boston cut, including an open cut for some distance on the crest of the ridge. No mining has been conducted here for some time, and a stamp mill that was formerly operated on the property has been removed.

IVEY MINE.—The Ivey mine is situated a short distance southwest of Dahlonega. The principal mining operations have been on lots 860 and 861, 12th district, a little to the east of Cane Creek.

Stover's Branch, a small stream flowing through the property, was the scene of active placer mining in the early forties at this locality. The placer deposits along this branch are said to have been unusually productive and have been worked over a number of times. Auriferous veins were prospected on lot 860, about 1879, and shortly afterwards the property came into the possession of the Consolidated Gold Mining Company. This company made a large cut about two hundred yards long in the saprolites at the junction of lots 860 and 861. A twenty-stamp mill was erected, which was replaced in 1884 by one of sixty stamps. A water ditch a number of miles in length was also constructed. No work of any importance seems to have been carried on since the above mentioned company operated the mine in the eighties. The stamp mill is now in an extremely dilapidated condition.

It having been so long since any mining has been done at the large cut, little can be learned about the character of the veins. It is stated in Bulletin 4-A, of the Geological Survey of Georgia, published in 1896, that a gold-bearing belt about eighty feet wide occurs in the schists at this locality, small included quartz stringers being the principal source of the gold. The results of two assays of ore samples from the Ivey cut are found in the publication just mentioned. One shows a value of $5.17 per ton and the other a value of $1.24 per ton. The property is owned at present by the Frank W. Hall estate.

CAPPS MINE.—The Capps mine on lot 890, 12th district, is a short distance from Dahlonega on the north side of Crown Mountain, a portion of Findley Ridge. A large open cut was made here a number of years ago and some prospect shafts sunk. Several auriferous quartz veins are said to occur at the locality, but no mining operations of any consequence have been carried on for a number of years.

BOWEN LOT.—This lot, No. 931, 12th district, lies close to the Capps mine. Some prospect work in the nature of tunnelling and sinking of shafts has been carried on on this lot in the past. No extensive work, however, has ever been done here.

FISH TRAP MINE.—This mine is situated on the south side of Crown Mountain not far from the two previously described properties. Several land lots are embraced in the property and before the Civil war the southernmost lots lying near the base of the mountain were the scene of placer mining. Vein and saprolite deposits have been worked on lots 932 and 964, 12th district. Some ore chutes reported as being quite rich were worked here in the early forties. After the Civil war the property changed hands several times in the course of its history and was worked for short periods by a number of different parties. No regular mining work has been carried on in recent years.

In the course of the various operations that have been conducted here several large cuts have been made in the saprolites and a number of shafts sunk and some tunnels driven. With the condition the old works were in when the property was visited no reliable information can be given in regard to the character of the deposits. The property is owned at present by the Frank W. Hall estate.

SKYRME MINE.—This mine is situated about three miles southeast of Dahlonega near Long Branch. Some vein mining was done here about fifteen years ago. A tunnel was driven into a hillside and a stamp mill erected. The mill and mill house are still to be seen on the property, but nothing could be learned concerning the character of the veins.

TEAL PROPERTY.—This property, on lot 122, 1st district, lies a short distance to the northeast of the Skyrme mine. A tunnel has been driven for a short distance and a little prospect work done on what is reported to be an auriferous saprolite zone. An interesting geological feature at this point is the occurrence of a well defined quartz vein of several feet in thickness, the strike of which is about at right angles to the trend of the rocks of the region. It is reported that this vein carries values where it is intersected by other veins or auriferous zones conforming with the schistosity of the country rocks.

CHESTATEE MINING PROPERTY.—This property consists of a number of land lots, lying principally along the Chestatee River, a few miles southeast of Dahlonega.

Some years ago a company did considerable work at this locality with the object of turning the course of the Chestatee River and working the river bed for gold. A hydraulic plant was erected on lot 145, 11th district, at the Chestatee bridge on the Middle Gainesville road. A large sum of money was expended, but the results not proving satisfactory the work was suspended and the property sold. At present it forms a part of the Frank W. Hall estate.

CROWN MOUNTAIN MINE.—Extensive mining operations have been carried on in the past in auriferous saprolites on Crown Mountain, a high knob of Findley Ridge. The location is about a half of a mile south of the Dahlonega public square.

Judge W. W. Murray, of Huntington, Tenn., formerly owned the property and under his directions three tunnels of considerable length were driven into the hill from different points. Part of these old workings were afterwards absorbed in open cut work done under a later management.

About eight years ago a company, of which General A. J. Warner, of Gainesville, Ga., was president, conducted some extensive hydraulic mining operations at this locality. A reservoir was constructed and water brought to the hill and some very large excavations made in the soft saprolite material. The deposits here consist of auriferous saprolites containing numerous stringers of quartz. As is usually the case with the majority of these auriferous zones, some larger, more well defined quartz veins than the average, are encountered. A flume line branching to the different excavations worked by this company conveyed the material cut down by hydraulic giants to the milling plant situated on a branch nearly a mile to the east of Crown Mountain. At the machine shops of the company on the Middle Gainesville road between the mines and the mill the heaviest, most resistant ore was separated out into bins and trammed to the mill. A very complete milling plant was erected by this company. The main mill consists of fifty 1,050-pound stamps and had in addition originally six Wilfley concentrating tables, but only three are in place at present. In this mill the more resistant ore was treated. In addition to the separation of the harder and softer ores that took place at the bins near the machine shops a log washer was used in connection with the flume line at the mill to complete the separation. The finer material was carried on beyond the log washer to another milling outfit where it was treated. This originally consisted of two Huntington mills, one of which was later replaced by a ten-stamp battery of 250-pound stamps. The milling plant was run by electrical power. No regular mining operations have been in progress for a number of years and the flume lines are in need of repairs.

The property is now owned by the Georgia Gold Mining and Power Company, Mr. Frank Moore, of Dahlonega, being president.

COLUMBIA MINE.—This mine, on lot 988, 12th district, lies just to the east of Crown Mountain mine. The first mining of any consequence done on this lot was in the early eighties. Some ore chutes reported to have been rich were worked at that time by Mr. W. K. Lawrence, then owner of the property. His work consisted of a large open cut near the top of Findley Ridge. After passing through several hands the property came into the possession of the company who did the extensive hydraulic work at the Crown Mountain mine. Some additional excavations were made at the locality in the nineties by this company. No mining operations have been carried on for a number of years and nothing definite can be stated concerning the character of the deposits. The property is owned at present by the Georgia Gold Mining & Power Company.

PREACHER MINE.—This mine is on Findley Ridge to the east of the last described property. A very large cut has been made here in the course of different hydraulic mining operations. Omitting minor workings, the first work of an extended character done at this mine was commenced in 1895 by Messrs. Murray, Moore, Clements and Harris, who erected a ten-stamp mill and began hydraulic mining, washing the material mined through a flume line to the mill. Later, the mine was operated by the same company who worked the Crown Mountain mine. Under their management the Preacher mill was removed to the large mill mentioned in connection with the Crown Mountain mine and the flume line from both mines connected. No work has been done at the Preacher mine for some years, and with the cut, as it is at present, a satisfactory description of the ore bodies could only be given by some one familiar with past conditions.

A fracture in the formation is noticeable near the upper end of the cut crossing it diagonally with a northwest trend. More or less quartz, some of it showing comb structure, has been deposited in this fracture. Mr. Joseph Clements, the veteran prospector of the Dahlonega region, is of the opinion that this fracture is continuous with the prominently out-cropping quartz vein that traverses the trend of the schists at the Teal property near Long Branch. The central portion of the lower end of the Preacher cut exhibits a considerable body of the hornblende series of rocks of the region ("Brick-bat" of the miners). The larger portion of the cut has been made in the saprolite of the acidic quartzose schists that are prominent along the course of Findley Ridge. The mine is owned by the Georgia Gold Mining and Power Company.

GRISCOM MINE.—This mine is on lot 996, 12th district. Its location is on Findley Ridge immediately adjacent to the Preacher cut. Extensive excavations have been made here by hydraulic mining. No mining operations have been carried on recently. The occurrence of the gold and the character of the deposits are said to be the same as at the Preacher mine. The Georgia Gold Mining and Power Company own this property.

BAST MINE.—This mine, on lot 1035, 12th district, is on Findley Ridge a short distance northeast of the Preacher and Griscom cuts and about one mile from the centre of Dahlonega. Hydraulic mining operations of an extensive character have been carried on here in the past together with some tunneling and sinking of shafts. A cut has been made in the saprolites in the course of different mining operations, comparable in size with the other large cuts about Dahlonega. In the early eighties the mine was operated by the Consolidated Gold Mining Company which at the same period was carrying on work at the Ivey mine. No work has been done at the Bast mine recently.

With the mine in its present condition, nothing very definite can be stated concerning the occurrence of the gold. Both acidic and the more basic hornblende series of rocks are exposed in the cut. The analysis of a specimen of the acid rocks taken from the southwest end of the Bast cut is given in the rock descriptions at the end of the chapter. It will be noted that the silica content is slightly over eighty-eight per cent. with a little more than eleven per cent. of iron oxide. The rock is made up practically of quartz with accessory magnetite. Some silicate of iron is probably present as is indicated by the amount of ferrous iron shown in the analysis. While no positive proof of original sedimentary origin, such as rounded grains of quartz with secondary silica rimes, can be obtained with the microscope, yet this rock suggests sedimentary origin more strongly than any other from the Dahlonega region described in the text.

DRY HOLLOW MINE.—This mine, on lot 126, is situated only a short distance from the south bank of the Chestatee River. Considerable work was done here years ago and the mine has the reputation of having yielded some rich ore. At the time the property was visited, Mr. Wm. McAfee was sinking a shaft on a vein at this locality, but on account of the presence of water no samples could be taken for assay. Since field work was completed, Mr. McAfee has exhibited some specimens of ore from this mine showing large amounts of free gold.

FINDLEY MINE.—This mine is situated at the northeast of Findley Ridge a little over a mile from the Dahlonega public square. The most important mining operations have been conducted on lot 1,048, 12th district. Two very large cuts have been made here by hydraulic mining. The lower or east cut commences at the foot of the ridge and is connected at its upper end with the upper or west cut situated near the top of the ridge. Plate III, Fig. 1, gives a view of a portion of the upper, or west cut.

This mine has received great notoriety from the occurrence near the south edge of the upper cut of a very rich ore chute from which a great deal of gold has been secured in the course of the history of the property. An inclined shaft was sunk on this chute before the Civil war and different owners of the property in later years have prosecuted underground work on the chute with varying success. It is stated by parties familiar with the history of the mine that something like $300,000 have been secured from this ore chute alone. No work has been done on the chute for some years and the old underground workings are inaccessible. Neither has any mining of an important character been carried on in the cuts for some time. When the property was visited, Mr. Thomas McDonald, of Dahlonega, was taking out some ore from the upper cut and milling it at the plant of a neighboring mine.

Owing to the amount of debris that has accumulated in the west or upper cut from rain-wash and caving while the mine has been idle the veins could not be examined with any degree of satisfaction. In Bulletin No. 4-A, of the Geological Survey of Georgia, published in 1896, three or four veins are mentioned as occurring in this cut. Two of them are spoken of as having a thickness of several feet.

The following assays of samples from these veins, together

with a description of the sampling are here quoted from the publication just mentioned.

"No. 1_____ 0.40 oz. ($8.26) of gold per ton.
"No. 2——— 0.50 oz. ($10.33) of gold per ton.
"No. 3_____ 0.45 oz. ($ 9.30) of gold per ton.
"No. 4_____ 0.55 oz. ($11.36) of gold per ton.
"No. 5_____ 1.10 oz. ($22.73) of gold per ton.

"Nos. 1 and 2 were taken from the vertical vein; the former, from the middle of the exposed part of the vein along its trend to the top of the wall on the west, and the latter, from the middle to the bottom of the open cut; No. 3 was from the Dead-Horse vein; No. 4, from the adjacent parallel vein; and No. 5, from the Cement vein." These samples were taken and the assays made by Dr. Thomas L. Watson, then Assistant State Geologist of Georgia.

The east cut has a steep slope from near its junction with the upper cut to the base of the ridge. The milling plant to which the material mined was flooded is situated not far distant by a creek on lot 1,087. The mill as originally constructed was a forty-stamp mill of 450-pound stamps. It is now badly in need of repairs.

In one wall of the east cut near its lower end an exposure of numerous stringer veins of quartz in the schist is to be seen. Plate III, figure 2, gives a view of this portion of the wall of the cut. It may be regarded as thoroughly typical of this class of deposits. Higher up the cut than the locality just spoken of, a large body of vein quartz is exposed. With the assistance of Mr. Thomas McDonald of Dahlonega, a sample for assay was taken from a section of this ore body about fifty by twenty-five feet. This sample on assay yielded no gold. The results, however, of this single assay test should not be taken as final, as this body of quartz is generally believed, by those acquainted with the mine, to be a low grade ore body. The property is owned by the Water Power and Mining Company of Georgia.

LOCKHART MINE.—The Lockhart mine, on lots 1,085 and 1,086, 12th district, is situated a short distance northeast of the Findley mine and a little over a mile from the Dahlonega public square. It is stated that the first regularly manufactured and equipped mill installed in the Dahlonega region was erected on this property and for the past twenty years important mining operations have been carried on more or less continuously. In addition to several large open cuts and a number of tunnels, a shaft has been sunk at the site of the present mill house near Yahoola Creek and a limited amount of drifting and stoping carried on. The important ore bodies here consist of stringers and lenses of quartz interlaminated with a dark green garnetiferous schist. As stated by Mr. Waldemar Lindgren[1], this garnetiferous rock associated with the vein quartz evidently represents an alteration by the vein depositing solutions of the normal surrounding mica schist. In addition to mica schist, a hornblende series of rocks is represented at this locality. Several observers have noted the occurrence at this mine of native gold in crystals of garnet.

As no work was in progress at the time the property was visited, the shaft could not be entered without putting the management to the expense of pumping out the mine, and no samples were taken for assay. From data gathered, however, from several geological publications and from reliable parties familiar with the mine, there seems to be reasonable grounds for the belief that there are here considerable bodies of ore of a good grade. A twenty-stamp mill of 450-pound stamps is located near Yahoola Creek.

The property is owned at present by the Frank W. Hall estate.

---

1. Bull. No. 293, U. S. Geol. Survey, p. 126.

Boly Field Mine.—This mine, on lot 1182, 12th district is situated on the north bank of the Chestatee River about two and one-half miles southeast of Dahlonega. In the forties, a small quartz vein was discovered at this locality immediately adjacent to the bank of the river, concerning which marvelous stories of rich yields of gold are told by parties familiar with the early history of gold mining in Lumpkin county. The vein was worked to a limited extent by Mr. Field, the original owner of the property at the time of discovery. Several attempts have been made since to mine this vein, but with indifferent success. No mining plant that could furnish power for a steam or air drill is located in the neighborhood and it is stated that the wall rock is so hard and tough that little progress can be made in operating a drill by hand. The gold is reported as occurring in chutes on the vein, and it is stated by reliable parties who worked in the mine in the forties that gold was secured on one occasion in such a large mass that it was cut up into convenient sized pieces on a blacksmith's anvil.

Beers Lot.—Both vein and placer mining has been conducted on this lot to a limited extent in the past. The locality is on the south side of the Chestatee River opposite the Boly Field mine. Mr. Joseph Clements, of Dahlonega, discovered and worked some rich pockets at this locality a number of years ago.

At the time the property was visited, Mr. Thomas McDonald and Mr. David A. Ritchie were conducting some placer operations of a limited extent in auriferous gravel deposits near the south bank of the Chestatee River.

Free Jim Mine.—The Free Jim mine is on lot 998, 12th district, in the town of Dahlonega. This mine was owned and worked in the forties by a negro man named James Boisclair.

Boisclair was not a slave and the mine became known as the Free Jim mine. In addition to the old workings of the original owner, some mining operations on a small scale have been carried on at the locality in subsequent years.

This mine seems to have attracted but little attention in recent years and on account of the lack of vein exposures nothing definite can be stated concerning the character of the deposits. The property is reported as being controlled by Mr. George H. Breyman, of Toledo, Ohio.

LAWRENCE, OR STREET, MINE.—This mine, on lot 951, 12th district, is in the town of Dahlonega a little north of the public square. Several shafts have been sunk at different points on an auriferous vein at this locality. No work has been done in recent years and some of the old shafts have been filled up in the course of municipal improvements. The remains of an old stamp mill are still to be seen at one of the principal shafts.

It is generally reported that some ore of a good grade was obtained from this mine, but nothing can be seen of the vein at present. Granite outcrops a short distance to the north of the old workings, while the hornblende series of rocks of the region is to be seen a short distance to the south.

HORNER MINE.—The Horner mine, on lot 855, 12th district, is situated about a mile to the northwest of the Dahlonega public square and near the northwest edge of the gold belt. This mine was first worked about twenty years ago by Mr. Joseph Clements and Dr. N. F. Howard, who erected a small stamp mill on the property. Later, the mine was operated for a couple of years by other parties who erected a larger stamp mill than the original one. No mining operations have been in progress for a number of years and nothing definite can be stated concerning the character of the deposits.

HAND, YAHOOLA, MARY HENRY AND BENNING MINES.—These several mines, all adjoining, are located near the town limits of Dahlonega on the eastern side. About 1901, the property of all of these, including several land lots in the 12th district, was consolidated into one property and the different mines operated by a company, known as the Consolidated Gold Mining Company. A 120-stamp mill was erected at the Hand mine on the site of the old Hand mill near Yahoola Creek. A smaller ten-stamp mill of 1,050-pound stamps was also installed at the Mary Henry mine higher up Yahoola Creek. A chlorination plant of fifty-ton per day capacity was built in connection with the large mill and some extensive tunnelling was done connecting some of the larger cuts. A shaft was sunk on a vein in the large Knight cut and a very large shaft-house built over it. On the east side of Yahoola Creek near the Mary Henry dam an inclined shaft was sunk for a depth of about sixty feet on a vein known as the Benning vein. The company just mentioned only operated these mines for a very limited period. They are now controlled by the Water Power and Mining Company of Georgia. No work has been in progress for several years. A large ditch known as the Hand ditch supplies the property with water for mining purposes.

Work on auriferous veins was carried on at the Hand mine as early as in the forties. Later, more extended operations were carried on by the Yahoola River and Cane Creek Hydraulic Hose Mining Company, and in the seventies the mine was worked by a company known as the Hand Gold Mining Company.

With the several large cuts that form the Hand mine in their present condition, little can be stated concerning the occurrence of the gold. In the large cut in which the shaft and shaft-house previously mentioned are located, a vein is said to occur known as the Knight vein. The results of an assay of a sample from this vein at a point where it showed a thickness of fifteen feet (a horse of mica schist five feet thick being omitted) is recorded in Bulletin No. 4-A, of the Geological Survey of Georgia, published in 1896. This assay showed a value of $11.88 per ton. In the same publication a vein at this locality is described, called the Antonio vein, from which a sample for assay was taken at a point where the vein was about twelve feet in thickness. This assay given in connection with a description of the veins shows a value of $5.68 per ton.

Two types of rock are to be seen on the dump at the main shaft in the Hand cut; one a dark colored mica schist composed principally of quartz and biotite and the other a light colored finer grained sheared granite. The latter, presumably, occurs as dikes in the schist, as dikes of a similar rock are to be seen at the contact of mica schist and granite a little to the northeast at the Benning mine. This granitic rock at the Hand cut consists, as seen under the miscroscope, of a fine grained aggregate of quartz and feldspar with some muscovite. A small amount of calcite and chlorite is also present. The rock, originally granitic in texture, is now distinctly schistose and the muscovite foils are probably of secondary origin. In polarized light the feldspar grains present a blotched appearance showing that they are not of uniform composition. The most of the feldspar is probably albite as is indicated by the chemical analysis which is given in another chapter.

The Benning shaft on the east side of Yahoola Creek was sunk on an incline to a depth of about sixty feet. The vein occurs here close to the contact of mica schist and a large body of granite. The mica schist appears from field examination to be identical with the schist at the Hand cut.

The Benning granite is a rather coarse grained, light colored granite showing to the unaided eye quartz, feldspar and biotite and some sericitic muscovite. The close similarity between the granite dikes occurring at the Hand cut and some noticed at the periphery of the Benning granite suggests an original magmatic connection.

STANDARD MINE.—The Standard mine lies to the east of the Hand and Yahoola mines on the opposite side of Yahoola Creek. Under this heading is embraced the mines formerly known as the Singleton mine and the Tahloneka mine. The latter, when a separate property, embraced lot 1,083, 12th district, the Singleton being on lot 1,084 and portions of adjacent lots.

The Singleton mine has a varied history, having been worked more or less continuously for many years. A very large cut has been made by hydraulic mining on the side of Singleton hill; the greater portion of this cut is on lot 1,084. With the cut in its present condition an examination of the deposits is very unsatisfactory. Three veins are stated to occur here. In Bulletin No. 4-A, of the Geological Survey of Georgia, published in 1896, six assays are given of samples from these three veins at different points. These assays are quoted below in numerical succession, it being impossible now to locate the exact localities in the cut from which the samples were taken:

No. 1_____$ 7.23 of gold per ton.
No. 2_____   49.61 of gold per ton.
No. 3_____ 261.99 of gold per ton.
No. 4_____   19.64 of gold per ton.
No. 5_____   4.13 of gold per ton.
No. 6_____   29.97 of gold per ton.

The Gowdy lot, 1,083, lies to the north of the Singleton

cut. A shaft was sunk on a vein on this lot about five years ago by a company who then owned the property. This shaft is equipped with a large shaft house similar to the one at the shaft in the Hand cut.

At the time the mine was visited, Mr. Wm. Campbell, of Dahlonega, was doing some work under a lease near the line of the Gowdy lot. Two inclined shafts were being sunk which, when examined, were each about thirty feet deep. A sample for assay was taken from a quartz vein in the bottom of each. At both localities the vein, where sampled, was about three feet in thickness, mainly solid quartz, but with some intercalated schist. The vein in the shafts presented a rolling, contorted appearance, varying considerably in thickness at different points and not giving very flattering prospects for a large, continuous supply of ore. The two samples yielded on assay $21.49 and $31.87 per ton respectively. The ore extracted was being milled at a stamp mill located near by on Yahoola Creek.

The rocks at the Standard mine present no features unusual to those described at neighboring localities. In the large Singleton cut, a light colored rock occurs that probably represents a sheared granite dike.

The Standard mine is controlled by the Standard Mining Company, Mr. George H. Breymann, president.

Jones Mine.—This mine, on lot 512, 15th district near the Chestatee River, is about four miles southeast of Dahlonega. Years ago, some work was done here on a vein that is reported to have yielded a large amount of gold. No mining has been in progress at the locality for many years and nothing can be seen of the vein.

At the time this property was visited, Mr. McAfee, of Dahlonega, was carrying on some prospect work on lot 512, not far from the old mine.

CAVENDER'S CREEK MINING PROPERTY.—This property, embracing several land lots in the 15th district, is on Cavender's Creek about five miles northeast of Dahlonega. In addition to some placer working in the past, a limited amount of underground work has been done on auriferous veins on two or three of the lots. About five years ago the owners of the property constructed a ditch a number of miles in length to bring water from Spencer's Creek and other streams, but for some reason little or no mining has been carried on in recent years. A small stamp mill is located on the property on Cavender's Creek. In Bulletin No. 4-A, of the Geological Survey of Georgia published in 1896, the results of assays of seven ore samples taken from this property are recorded. These assays are given below, together with the localities as there stated:

No. 1____0.23 oz. ($ 5.17) of gold per ton.
No. 2____0.99 oz. ($20.46) of gold per ton.
No. 3____0.20 oz. ($ 4.13) of gold per ton.
No. 4____1.27 oz. ($26.25) of gold per ton.
No. 5____0.92 oz. ($19.02) of gold per ton.
No. 6____0.80 oz. ($16.54) of gold per ton.
No. 7____1.07 oz. ($22.12) of gold per ton.

No. 1 was from veins in a tunnel on the west side of lot 390; No. 2, from a second vein on the same lot; No.4, from a vein near the southwest corner of lot 373; No. 5, from two veins at the top of the hill on lot 389; No. 6, from a forty-foot stringer lead at the bottom of the same hill on lot 390; No. 7, from a four-foot vein near the northeast corner of the last mentioned lot.

The property is owned at present by the Cavender's Creek Gold Mining Company of Dahlonega.

JUMBO MINE.—The Jumbo mine, on lot 374, 15th district,

is about six miles northeast of Dahlonega. Several auriferous quartz veins have been prospected at this locality sufficiently to give some idea of their values. In addition to the veins on which work has been done, it is claimed that there occurs here a broad zone of saprolites that carries values. Two double compartment shafts, about seventy feet apart, have been sunk and connected by a drift at thirty feet from the surface. These shafts could not be entered at the time the property was visited on account of water, and nothing can be stated concerning the character of the vein. At the northeasternmost of these shafts a sample for assay was taken from an ore pile of several tons. This sample yielded on assay $6.89 per ton.

On the top of a hill in the vicinity of these shafts a large quartz vein has been exposed by a tunnel for several rods. A sample for assay was taken from sections across this vein at three points. This sample on assay yielded $2.06 per ton. This vein, where sampled, showed an average width of seven or eight feet. On a hill about two hundred yards northeast of this locality two small auriferous stringers have been located, designated by the owners of the property as Mistletoe vein No. 1 and Mistletoe vein No. 2. These had not been sufficiently exploited up to the time of examination to warrant a definite description. A ten-stamp mill has recently been installed at the mine.

The property is owned by the Jumbo Gold Mining Company, Mr. J. F. Moore, president, and Mr. Joseph Clements, assistant general manager.

The mining operations carried on here, so far, have been conducted economically and on a small scale with the aim of exposing the deposits. Work of this class cannot be too strongly commended for its beneficial influence on the future of the gold industry as compared with some extravagant expenditures that have been made in the Dahlonega belt.

GARNET MINE.—This mine, in the 15th district, is a short distance northeast of the Jumbo mine and about seven miles from Dahlonega. Placer work was carried on along branches on the property years ago. The first vein mining of importance was commenced in the eighties by the Garnet Water Power and Mining Company. Considerable hydraulicking was carried on in auriferous saprolites containing quartz stringers. A twenty-stamp mill was erected and operated by water power from a dam across the Chestatee River. Recently some hydraulic mining has been done on the property about a fourth of a mile southwest of the old cuts. A considerable excavation has been made in the saprolites and a mill of ten stamps, taken from the old mill has been erected several hundred yards from the cut on a small branch. The ore is flooded from the cut to the mill through a flume line. No work was in progress when the mine was visited and nothing can be stated concerning the character of the deposits.

OTHER PROPERTIES IN LUMPKIN COUNTY.—In addition to the properties already described, a number of localities are to be noted where either regular mining has been carried on to a limited extent or prospecting done. Some mining has been done on an auriferous vein on lot 955 and vicinity, 12th district, a short distance north of Dahlonega. A ten-stamp mill was erected on Yahoola Creek when the work was in progress.

THE RIDER MINE is on lot 1,058, 12th district, northeast of Dahlonega. Vein mining on a small scale was carried on here years ago.

Mr. E. E. Crisson was conducting some hydraulic operations in saprolite material a short distance northeast of Dahlonega at the time the region was visited. The work was being prosecuted near a vein known as the Hamilton vein, on which some shafts had been sunk and tunnels run years ago. Mr. Crisson was milling the ore sluiced from the cut in a small stamp mill of home manufacture.

THE CORA LEE PROPERTY is on lot 433, 15th district, near the Cavender's Creek property. Some prospect work on veins has been done here in the past by Mr. Joseph Clements, of Dahlonega, and others.

THE BLUFFINGTON MINE is near the Jones mine, southeast of Dahlonega. Work was carried on here many years ago that is reported to have been profitable.

THE OLD COLUMBIA MINE is south of the Findley mine. Work was done here years ago by a company known as the Columbia Mining Company.

THE WOODWARD LOT lies to the southwest of the McAfee-Lynn mine. A limited amount of hydraulic work has been done here in the saprolites in the past. A small stamp mill is still on the property.

THE KEYSTONE MINE is on Cane Creek below the Barlow mine mill. Some auriferous veins have been worked here years ago.

THE STEGALL PLACER, the Belle mine, the Saltonstall mine and the Woods mine are all located near Auraria. They were operated for a short time some years ago.

THE WELLS MINE, on lot 1,213, 12th district, is about a mile and a half southwest of Auraria. An auriferous vein was worked here and a small stamp mill erected.

THE HIGHTOWER MINE is situated near the Wells mine. It was operated for a short time in the eighties.

Mining operations have also been carried on to a limited extent on lots 891, 930 and 725 in the 12th district.

DREDGES.—Several dredges have been operated at different times on the Chestatee River. Some of these are said to have paid very well. The boats of Messrs. Birch Brothers, of Kansas City, and Mr. H. D. Jaquish, now of Gainesville, Ga., are reported to have been very successfully operated. At the time the region was visited, a dredge was in process of equipment for the purpose of securing auriferous black sands from the bed of the Chestatee River.

WHITE COUNTY

In White county, the Dahlonega belt has about the same average width as in Lumpkin, except at the extreme eastern edge where it narrows before entering Habersham county. The most extensive mining operations have been carried on in the region about Nacoochee Valley in the eastern side of the county. White county gained notoriety years ago when placer mining was being actively prosecuted by reason of the number of large nuggets that were obtained from the Nacoochee Valley region, some of them weighing as high as five hundred pennyweights. Numerous nuggets of considerable size have also been found at the Loud mine in the western part of the county. While considerable vein mining has been done, and some large excavations made at several localities in mining saprolite deposits, yet, near many of the more important placer deposits, prospecting for veins does not appear to have been carried on very vigorously. This county offers an interesting field for future vein prospecting.

LOUD MINE[1].—This mine, on lots 39 and 40, 1st district, is about four miles southwest of Cleveland, the county seat of White county. An area of a number of acres of alluvial

---

1. Special thanks are due Mr. R. K. Reaves, Jr., manager, and his foreman, Mr. Gabriel Furgerson, for courtesies extended during field work at this mine and neighboring properties.

deposits with underlying auriferous gravels occurs here in a stretch of land traversed by a small branch and having very slight drainage fall. The area forms a prong of the extensive lowlands along Town Creek and extends back from where these occur on the eastern side of lot 39 to near a low divide on lot 40 known as Hog Back Ridge. A great deal of placer gold has been obtained in the course of different placer mining operations at this locality. Some nuggets have been found weighing as high as three hundred pennyweights or more. Although this mine has been worked at different times through a period of many years, yet, owing to slight drainage fall, the deposit has never been systematically worked as a whole. Mining operations have been carried on by working the gravel deposits in pits at different points with small hydraulic lifts. As the overburden is rather heavy at some localities, probably as much as twenty feet in thickness, this method of mining has proved very unsatisfactory, owing to the inconvenience arising from the accumulation of the tailings. Much of the area has been worked over; some of it several times. In the case of flat lying placer deposits, that have been worked as this one has been at different times through a long period of years, it is difficult to reach satisfactory conclusions as to what amount of gold still remains for future mining operations. Owing to the richness of this deposit and the difficulties attendant on mining it thoroughly, as above stated, it is probable that a considerable amount of gold could yet be obtained if good drainage was secured and the deposit was worked systematically and exhaustively. With this end in view, several companies have, at different times, prosecuted work in an effort to drain the area from its upper end, but their plans were only partially carried out. As the inauguration of these efforts, the Hand Mining Company, a number of years ago, made a large cut on lot 66 to the southwest of the mine near the Tesnatee River. The plan was to establish a water way from the Tesnatee River to the Loud mine, reversing the natural drainage of the lowland in which the deposits occur. Advantage was to be taken of stream valleys as far as possible from the river to Hog Back Ridge, the divide previously mentioned as occurring near the head of the Loud Valley. At this locality a tunnel through the ridge into the Loud Valley was contemplated. The Canadian-American Loud Gold Mining Company, who recently controlled the property for a short period, extended the drain above referred to to the southeastern corner of lot 65.

In addition to the placer deposits at the Loud mine, a very interesting auriferous vein occurrence merits description. A number of years ago the outcrop of a small vein was located in the placer deposit. A shaft that had been sunk on this vein years ago was deepened about five years back and drifts run northeast and southwest for approximately thirty feet by Mr. R. K. Reaves, of Athens, Ga., the present owner of the mine. Some wonderfully rich chutes or pockets were struck in this vein and magnificent specimens of crystallized and wire gold obtained. Inability to cope with the water is stated as the reason for only a limited amount of work having been done on the vein. The operations carried on were not sufficiently extensive to furnish definite data as to the pitch of the ore chutes and the probable frequency or regularity of their occurrence. Some very coarse gold obtained in panning a little of the old debris about the mouth of this shaft gave testimony to the richness of some of the ore that had been taken out. Something like two hundred yards from the shaft above mentioned, two other shafts near each other were sunk in the eighties to a depth of about seventy feet. It is reported by Mr. Gabriel Furgerson, foreman of the mine, that about a thousand dollars worth of gold was secured here. As the different shafts are in line with the general strike of the formation, it is probable that they are on closely associated veins, if not on the same vein. Judging from the fact that much of the gold that has been found in the placer presented features similar to the gold in the vein or veins above described, and also from the failure of prospectors to locate any notable deposits in the neighboring slopes, it seems certain that a portion, at least, of the placer gold has been derived from veins occurring in the rocks beneath the alluvial deposits. This supposition is further strengthened by the outcrop of a body of hornblendic rock along the southaest side of the Loud Valley associated with more acidic schists and gneisses. The occurrence, as frequently mentioned in previous pages, of many of the most important gold deposits, that have been so far located in the Dahlonega belt,at or near the contact of hornblende schists with more acidic rocks is too significant to overlook in considering the probable location of auriferous veins. Plate II, Fig. 1, shows a photograph of a specimen of gold crystals, associated with quartz crystals, from the Loud mine. In addition to the vein deposits already described at the Loud mine, an open cut about fifty yards in length was made several years ago on an auriferous vein or zone at the base of the slope on the southeast side of the Loud Valley. The ore was milled at a ten-stamp mill which is still on the property in the lowland near the cut. Some hurried panning tests in this cut gave unsatisfactory results, but as no work had been in progress for several years the deposit was not very fully exposed.

Southwest of the Loud Valley, considerable placer areas are found on lot 56 along two or three small streams. These are known as the Asbury placers, having been rather extensively worked in former years by a gentleman of that name.

FIG. 1.—MINING PLANT, SEMINOLE GOLD AND COPPER MINE, LINCOLN COUNTY, GEORGIA

FIG. 2.—MILLING PLANT, COLUMBIA GOLD MINE, McDUFFIE COUNTY, GEORGIA

At the time the property was visited, some placer work was in progress on this lot under the direction of Mr. R. K. Reaves, Jr. These operations, on portions of the Asbury placers that had never been mined were being carried on with a hydraulic lift.

West of the locality just mentioned along a ridge on lots 65 and 57, a line of old workings are to be found where mining operations were conducted on an auriferous vein many years ago. The ore bodies were not exposed, but some panning made from the material of a dump at the mouth of one of the old shafts yielded a fair amount of gold. Old works of a similar character to the ones at this particular locality are to be found throughout the entire length of the Dahlonega belt. As in the early mining period, veins carrying good values were sometimes abandoned after they had been mined down to water level, either from the ore ceasing to be free milling or from inability to cope with water, these abandoned mines offer a good field for future tests.

Courtney Placer.—On lot 33, southeast of the Loud mines, placer deposits of considerable extent have been worked by Mr. Courtney, of Cleveland, Ga. A large cut was made in the course of this work extending from a point on a small branch at the edge of the extensive lowlands of Town Creek and the Tesnatee River westward past the old Courtney homestead almost to the house of Mr. Gabriel Furgerson. Fringes of unworked auriferous gravel are to be found at several localities along the edge of the worked area. From some of these unworked gravels considerable rather coarse gold was obtained in panning tests. Mr. Courtney states that near the Courtney homestead, at a point where a small storehouse now stands, unusually coarse gold was found and a tunnel was driven along the gravel deposit to one side of the open cut in mining for it.

Owing to heavy overburdens, only a limited amount of placer mining has been done in the large lowlands of Town Creek and the Tesnatee River. At a point on the former stream, known as Town Ford, a small area was worked years ago, and from there up to the Loud mine more or less mining is said to have been carried on along the borders of Town Creek. On the west side of the Tesnatee River, immediately above the public bridge a few acres of the main lowlands have also been mined.

Town Creek and Tesnatee River unite on lot 33 and the lowlands of these two streams on this lot and on the Henderson lot adjoining it on the north, together with the lowland along the first named stream on the Loud mine lot, united, form an extensive body of valley land, the larger portion of which has never been mined. This area as a whole may be a good field for future dredging operations. Systematic preliminary testing as to the amount of gold and the character of the bed-rock would involve considerable labor, owing to the thickness of the overburden.

HENDERSON PROPERTY.—The Henderson property, on lot 34, lies to the north of the Courtney lot and east of the Loud mine. Both vein and placer deposits have been located on this property, but very little development work has been undertaken so far. Extensive lowlands occur here along Town Creek and the Tesnatee River. These valley lands have just been spoken of in connection with the lowlands of the Courtney property as being a possible field for future dredging operations. Near some old placer workings on a small branch that flows through the extensive Tesnatee River lowlands several panning tests were made in an unworked portion of a gravel deposit on the slope of a hill adjacent to the valley lands just mentioned. A small pit was sunk to bed-rock a few yards from a tenement house at the locality and very satisfactory results obtained, the gold secured being quite coarse. It is not probable that there is any extensive area of gravels at this point. The tests, however, of gravels near small streams, both on this property and on the Courtney lot, are of interest as giving some indications of the probable amount of gold in the extensive lowlands of the two principal streams previously mentioned.

In addition to the placer deposits on the Henderson property, several prospect shafts and pits have been sunk on auriferous veins occurring in a ridge in the northeast part of the lot. No work had been done at the locality for some years and very little could be seen of the veins. Mr. Albert Henderson, of Cleveland, Ga., the owner of the property, states that in one of the shafts a small stringer or aurifeous band was exposed that yielded very rich pannings. Near a spring branch on the northeast side of the ridge the ore of a pile of vein quartz, taken out in the course of some prospect work, showed more or less crystal form. Both here and at the Loud mine a distinct tendency to crystal form in the vein quartz seems to be of common occurrence.

ETRIES PROPERTY.—On lot 62, 1st district, placer deposits occur along a branch that flows through the property. Lack of exposures prevented any very satisfactory tests being made at this locality. A little to the south, along a smaller tributary stream, old placer works were examined and some of the worked gravels yielded considerable gold when panned. On lot 59, adjoining lot 62, a little hydraulic work was carried on in saprolite material some years ago on a vein deposit. Several quartz veins outcrop on this lot. A small vein, or auriferous band, a few inches thick is exposed in several old shafts or pits in saprolite material. The vein material from one of these shafts yielded considerable quantities of gold when panned. On the lot line between lots 58 and 59 a quartz vein about five feet thick is exposed in a shallow pit. A sample for assay was taken from a section across this vein as exposed in the pit. The ore looked rather promising, but no gold was obtained in the assay.

MATTHEWS LOT.—On this lot, No. 49, 4th district, old placer works are to be found along a stream traversing the lot. Some underground work was also carried on years ago on vein deposits in a hillside near the placer. Some panning tests were made in the gravels along the stream above mentioned, but the results were not very satisfactory.

LOT 48.—On lot 48, 4th district, some placer work has been done in the past on this lot along a small stream known as Gold Branch. A little prospect work has also been carried on on veins.

LOTS 37 AND 38.—A limited amount of vein mining was done on lots 37 and 38 a number of years ago by Mr. R. K. Reaves, of Athens, Ga. On the former lot a tunnel was run for a short distance into a hillside along the strike of a quartz vein. The ore obtained in these mining operations was milled at a small stamp mill that was located on Jennings Creek some little distance from the locality where the mining was carried on.

McAFEE PROPERTY.—On lot 36, to the east of lot 37, a number of old shafts are to be found that were sunk years ago on what is supposed to be the continuation of the same vein that was worked on lot 37. Some placer deposits along small streams flowing through the lot have also been worked in the past. Some additional mining operations have been carried on on lot 25, 4th district.

SPRAGUE, OR BLAKE, MINE.—This mine, on lot 26, 4th district, is a few miles northwest of Cleveland, the county seat.

Considerable work was done here years ago on an auriferous quartz vein known as the Sprague vein. At the time the property was visited little, if anything, could be seen of the vein, as no work had been in progress for many years. In Bulletin No. 4-A, of the Geological Survey of Georgia, published in 1896, it is stated that the vein, exposed at that time in two shafts on lot 26, showed a thickness of four feet and contained large amounts of pyrites. According to local reports some very rich ore has been obtained from portions of this vein. The remnants of an old stamp mill are still to be seen by a creek that flows through the property.

Judging from the amount of work that has once been done at the locality together with the favorable reports of the character of the vein, this mine would seem to merit more attention than it has received in recent years. The property is in the charge of Mr. J. W. H. Underwood, of Cleveland, Ga.

Longstreet Mine.—This mine is situated a few miles north of Cleveland, the county seat of White county. The principal mining operations have been conducted on lot 162 in the 3rd district. One or two adjoining lots on which placer deposits have been located are also included in the property.

On lot 162 an auriferous quartz vein has been exposed by open cut work and a tunnel. The open cut is over two hundred feet long and at its northeast end a tunnel, reported to extend for approximately three hundred feet, has been driven into the hillside on the strike of the vein. The underground workings were inaccessible at the time the property was visited. The vein is exposed in the open cut southwestward from the mouth of the tunnel for about two hundred feet. As seen here it shows an average width of something like six feet and is composed of quartz with interlaminated bands of gneiss, or mica schist. The strike of the vein is about N. 45° E. with a nearly vertical dip. A few yards to the south, yellowish brown saprolite, most probably derived from a hornblende schist, shows in the flume-cut leading to the mill. Three samples for assay were taken from sections across the vein in the bottom of the open cut. No. 1 was taken at a point about midway between the mouth of the tunnel and the southwest end of the open cut; No. 2, at a point about seventy-five feet northeast of No. 1; and No. 3, from a point about fifty feet northeast of No. 2. These samples yielded on assay the following results:

No. 1_____ $2.06 per ton.
No. 2_____ 0.62 per ton.
No. 3_____ 0.41 per ton.

In the light of the work that has been carried on here, together with the amount of gold that was obtained in panning tests at various points on the vein and statements made concerning this mine, these assay results are disappointing. It should be borne in mind, however, that only tentative conclusions concerning the average value of a large body of ore should be drawn from such a limited number of assays.

A short distance northwest of the above described vein some open cut work has been done in saprolite material. In this cut a small vein or auriferous band yielded some very good panning results. From the open cut on the main vein a flume line leads to the mill situated about twelve hundred feet to the southwest. The milling plant is equipped with a Huntington mill and a Wilfley concentrating table.

In addition to the vein deposits, placer deposits occur along Turner's Creek on this property and also on some smaller tributary streams. Portions of these placer deposits have been worked by various parties at different times. At a point on the above mentioned stream on lot 162 where some unworked gravel was exposed in a sharp curve, some very satisfactory panning tests were made, the gold obtained being rather coarse. At the time the property was visited Mr. W. A. Danforth, general manager, was carrying on some placer mining in deposits in a small hollow or gulch on lot 162. Attention was called at this locality to a zone about two and a half feet wide of saprolite material that yielded a moderate amount of fine gold when crushed and panned. A small quartz stringer occurring in this zone yielded no gold from panning tests. The deep yellowish brown color of the saprolite admitted of little doubt that its derivation was from one of the hornblende rocks characteristic of the Dahlonega belt. This occurrence of gold is of interest, as several writers on the deposits of the Dahlonega region have reported that tests of the saprolites of that locality showed the portion of the gold which apparently was not in quartz occurred principally in acidic schists and gneisses.

LEWIS MINE.—This mine is situated about seven miles from Nacoochee Valley near the northwest edge of the Dahlonega belt. Some vein mining was done here years ago. Old shafts and open cuts are still to be seen near the crest of a small ridge. No mining operations have been carried on for many years and nothing definite can be stated concerning the character of the deposit.

BELL PROPERTY.—Placer deposits occur on lot 132, 3rd district. A considerable tract of valley land is on this property and old placer workings are to be seen on either side of a creek flowing through it. This mining done in the lowlands was carried on years ago and it is difficult to say how much of the deposit has been worked. Some hydraulic surface work has also been done in the past near the base of a hill on the west side of the valley. Pannings of detrital material from the surface of these old works yielded a little gold.

Cox AND MERRITT PROPERTY.—The creek flowing through the Bell property also traverses this property which lies immediately to the south. A considerable body of valley land occurring here is a continuation of the lowlands of the Bell property.

CASTLEBERRY PROPERTY.—This property adjoins the Bell property on the north. Some prospect work has recently been done here near the line of lots 131 and 132. This work consisted principally of a couple of tunnels run for some distance into a hillside. As no mining operations were in progress when the property was visited it was difficult to form a judgment as to the character of the deposits. In addition to this work recently done, several old tunnels were noticed in the vicinity. Some placer mining has also been carried on in the past at this locality along the same creek that flows through the Bell property.

THURMAN PROPERTY.—A limited amount of vein mining has recently been carried on on this property, lot 102, 3rd district. Several shafts have been sunk and a couple of tunnels run into a hillside. No work was in progress at the time the property was visited and as the entrance to both tunnels was closed they could not be entered. Nothing can be stated concerning the character of the deposit, but, judging from the appearance of a pile of ore near the mouth of one of the tunnels, a vein of considerable size probably occurs at the locality. Placer deposits are said to be found along a small stream flowing through the property.

WHITE COUNTY, OR THOMPSON, MINE.—This mine on lot 102, 3dd district, is situated about two miles southwest of Nacoochee post office. Considerable underground work has been prosecuted here at different periods on an auriferous vein consisting of quartz with intercalated bands of country rock. Near the top of a low ridge, a shaft, with a hoisting engine and shaft house, is located on the vein, but as no mining operations were in progress at the time the property was visited the underground works could not be entered. Southwest of the shaft, old works are to be seen along the strike of the vein for a considerable distance. Along the line of these old works the vein seems to have been pretty completely stoped out to the surface. At one point an exposure was noticed, but it was inaccessible for sampling. At this exposure the vein has a thickness of about four feet and shows a rather strikingly banded appearance due to the interlamination of country rock with the vein quartz. In Bulletin No. 4-A, of the Geological Survey of Georgia, published in 1896, the following statement is noted in regard to the value of the ore at this mine: "The average ore is low grade, probably varying from fifty cents (the assay made by the survey of this sample) to $5.00 per ton. * * * * The picked ore runs high."

In addition to the principal vein, a smaller vein has been exposed in a tunnel driven into the hill on its southwest side. Until further developments are prosecuted little can be stated concerning the character of this vein.

A milling plant is located on a small creek on the property. This is equipped with a ten-stamp mill, engine, boiler, etc., housed in a substantial building. The property is controlled by Mr. Herman Dye, of Detroit, Mich.

REYNOLDS AND HAMBY ESTATE MORTGAGE COMPANY'S PROPERTY.—This company controls a number of land lots' in the 3rd district near Nacoochee Valley on which gold mining has been prosecuted at different times. The more important mining operations have been confined to the region about Duke's Creek and the Chattahoochee River. Much of this property was formerly known as the Martin property, having been previously owned and worked by Mr. John Martin, now of Clarkesville, Ga. Both vein and placer mining have been carried on at several different localities on the property. Some very productive placer deposits were worked along branches emptying into Dukes Creek. Near the line between lots 70 and 71, on a prong of Black Branch, a nugget weighing about five hundred pennyweights was obtained in these workings.

On lot 70, at a point a short distance north of Dukes Creek, both hydraulic and underground mining was carried on a number of years ago on an auriferous vein known as the Reynolds vein. The first mining conducted here was of a hydraulic nature and a large cut or rather two cuts, which were later united, were made along the strike of the vein, which has a southwest-northeast trend. The most important mining carried on was done during the period from 1896 to 1902, the greater portion of the work being conducted by Mr. John Martin. In addition to some hydraulic operations, in the course of which the two cuts above mentioned were united forming the present cut which is about two hundred yards long, a shaft was sunk on the vein in the bottom of the cut to a depth of about a hundred and forty feet and considerable drifting done. The principal drifting was at the sixty-foot level. The longest drift was to the southwest and, combined with a tunnel, which had been driven on the vein from near Dukes Creek, had an approximate length of four hundred and fifty feet. To the northeast a drift was run for about one hundred and fifty feet. The ore was milled at a twenty-stamp mill erected near by. This mill was later reduced to one of ten stamps and at the time the property was visited it was badly in need of repairs. Mr. Martin estimates that during the period above mentioned between forty and fifty thousand dollars worth of gold was obtained. No work has been done at the locality for a number of years and the underground works were not accessible at time of visit. The vein is said to average several feet in thickness and to consist of quartz with more or less interlaminated wall rock.

On a portion of lot 60, northwest of the Reynolds vein, considerable mining was carried on a number of years ago. In addition to placer work, several shafts were sunk and other prospect work done on a vein at a point a short distance from the site of a building that was erected at the time as an office and laboratory. Nothing very definite can be stated concerning the character of the deposit here. At the time of visit, Mr. J. A. Bramlet was conducting some sluicing operations near by on the headwaters of Hamby Branch. A small stamp mill erected about four years ago is located on the above mentioned branch at this locality.

A short distance northeast of the Reynolds and Hamby mines and immediately west of the Chattahoochee River, another portion of the tract, formerly known as the St. George property, merits notice. In addition to considerable placer mining on deposits along the England, the Old House and the Gaten branches, a large excavation, known as the Dean cut, was made by hydraulic mining on an auriferous vein or zone on lot 38. No regular mining operations have been carried on here for a number of years and nothing definite can be stated concerning the character of the deposits. At the time of visit some sluicing operations of a limited character were being conducted along one of the branches above mentioned.

YONAH LAND AND MINING COMPANY'S PROPERTY, OR CALHOUN MINE.—This property embraces a number of land lots in the 3rd district contiguous to the property of the Reynolds and Hamby Estate Mortgage Company. Very productive placer deposits have been mined in the past along branches on this property emptying into Dukes Creek. Some vein mining has also been carried on. One of the most noted placer deposits of the Nacoochee Valley region was worked a number of years ago along Richardson Branch on this tract at a point a few hundred yards from Dukes Creek and close to the old Lumsden residence. The gold obtained was unusually coarse, much of it occurring as nuggets of several pennyweights and larger. As the drainage area of this branch is rather restricted, an inviting field is afforded at this locality for vein prospecting. Considerable placer mining has also been carried on in the valley lands along Dukes Creek below the mouth of Richardson Branch. At the time of visit a dredge boat that had been operating on the Chattahoochee River and Dukes Creek a year or so back was located in these lowlands.

Northwest of the Reynolds mine on lot 68, more or less vein mining has been carried on in the past. In Bulletin No. 4-A, of the Geological Survey of Georgia, published in 1896, the results of two assays are given of ore from a vein on this lot designated as vein No. 2. The vein was about a foot in thickness at the point where the samples were taken. These two assays showed values of $9.40 and $13.00 per ton respectively. At the time of visit a limited amount of vein mining had recently been in progress near the mouth of Mercer Branch.

CONLEY MINE.—A number of years ago a cut was made by hydraulic mining on lot 39, 3rd district. A little very fine gold was panned from granular quartzose material occurring at this locality, but with the cut in its present condition little can be stated concerning the deposit.

ROBERT'S PROPERTY.—Several land lots in the 3rd district are embraced in this property. Considerable mining operations, principally on placer deposits along small branches, was carried on at this property a number of years ago by Mr. Charles Roberts, the former owner. In addition to placer mining a small cut was made by hydraulic work on lot 26 on a rich pocket occurring in saprolite material. It is reported that after a limited amount of work this pocket was exhausted and no further remunerative returns could be obtained at the locality.

HARDEMAN PROPERTY.—An extensive body of lowland along the Chattahoochee River is embraced in this property which is in Nacoochee Valley near Nacoochee post office. Owing to heavy overburden, little or no mining has ever been carried on in the placer deposits here, though it is reported that nuggets of considerable size have been secured from the bed of the Chattahoochee River at this locality. As these valley lands lie near the southwest edge of what has, so far, proved the most important part of the Dahlonega belt in White county and as numerous tributary stream valleys a short distance above have afforded remunerative placer deposits, there is the possibility of there being here a good field for dredging operations.

PLATTSBURG MINE.—This mine on lot 40, 3rd district, lies just east of the Chattahoochee River in the Nacoochee Valley region. The mine is situated on a ridge rising abruptly from the river bank. Vein mining has been conducted here on an auriferous zone said to be from six to eight feet in thickness and composed of quartz stringers and intercalated wall rock. Several tunnels have been driven into the hill and some shafts sunk and open cuts made. No mining operations have been in progress for a number of years and with the mine in its present condition nothing additional can be added to the statement already made concerning the character of the deposits. From some mill tests of the ore, made shortly before Bulletin

No. 4-A, of the Geological Survey of Georgia was published
in 1896, the conclusions drawn and stated in that report are
that the ore is low grade, probably not exceeding three dollars
per ton.

A ten-stamp mill with two Frue vanner concentrators is
located contiguous to the mine on the bank of the river.
Messrs. Joseph Gruber and Wm. Voigt, Jr., are the principal
owners of the property.

FRANKLIN AND GLENN LOT.—This lot, No. 41, 3rd district,
lies northeast of the Plattsburg property. Some prospect
work has been done here, principally near the crest of a high
hill, on an auriferous zone consisting of country rock with
interlaminated quartz stringers. The exposures seen at time
of visit were hardly sufficient to permit satisfactory conclusions to be drawn as to the width of the zone and the probable
presence of other parallel zones. In a shallow pit near the
crest of the hill, the ore body showed an approximate thickness
of eighteen feet. A sample for assay was taken from a section
across this pit. This sample yielded on assay $0.82. A single
assay test at one point only is very unsatisfactory from which
to draw conclusions, yet, as the decomposition of the country
rock at the locality doubtless extends to considerable depths,
should future prospecting prove the presence here of a gold-
bearing zone of considerable dimensions, profits might be
made with hydraulic mining although the average values
might be no greater than the figures given above. The greatest drawback to future operations on the crest of the hill where
the zone is exposed is the absence of water, the altitude of this
particular point being considerably above any water in the
neighborhood. On the side of the hill in the vicinity of the
exposure above mentioned two tunnels a hundred feet or
more in length have been driven into the hillside, but the

results of this work seem to have thrown very little light on the nature and extent of the deposits.

In an open cut by a branch at the base of a hill an auriferous vein has been exposed on this lot. Owing to its position in a depression considerable material has accumulated in this cut. From some ore at its edge, and from what exposure is visible, it is judged that a vein several feet in thickness, composed of quartz with interlaminated mica schist, occurs here. Panning tests of some of the ore showed the presence of considerable gold. The property is owned jointly by Mrs. C. L. Franklin and Mr. James Glenn.

CHILDS MINE.—This mine lies a short distance to the northeast of the Franklin-Glenn lot. Several land lots are embraced in the property, but the most important mining operations have been conducted on lot 23, 3d district. A large cut has been made here in the saprolites by hydraulic work in the course of different mining operations. Underground mining has also been done from the bottom of the cut. The material from the cut was flooded through a long sluice line to a stamp mill situated on Bean Creek near the Jones mine. A very large amount of material was treated in this manner. No work has been done here for a number of years and with the cut in its present condition little can be stated concerning the character of the deposits. There are good reasons to suppose, however, that the auriferous deposit here is the southwest continuation of a gold-bearing zone of considerable width that has been mined rather extensively a short distance to the northeast at the Jones Mine, on lot 10. Mica schist is the prevailing rock at the Childs mine, (Chapter IV, description of rocks) but some bodies of saprolite that would seem from their color to have been derived from hornblende schist were noticed. In addition to the main cut above mentioned, prospect work has been done on auriferous veins, or gold-bearing zones, at a number of points on the property. On lot 24, near the line of lot 41 and southwest from the principal cut, an auriferous quartz vein about two feet in thickness has been exposed for a short distance in an open cut. Panning tests of some of the ore yielded rather unsatisfactory results. Several other veins have also been exposed to a limited extent at different localities on the property, but nothing very definite can be stated as to their probable value. Placer deposits along small streams flowing through the property have also been mined in the past.

The milling plant is situated, as stated above, on Bean Creek near the Jones mine, a considerable distance to the northeast of the large cut. The mill is equipped with forty-five stamps, and is operated by water power; water from a water ditch on a neighboring hill that supplies the mine with water for general mining purposes, being utilized.

It is regretted that complete statistics of the gold secured here in the course of different mining operations were not available, as it is generally reported that a large amount of gold has been obtained at this locality.

At the time of visit, the property was owned jointly by Chancelor D. C. Barrow, of the University of Georgia, and the heirs of Otis and A. K. Childs.

JONES MINE, OR LOT 10.—This mine lies immediately to the northeast of the Childs mine. A gold-bearing zone has been mined for a number of years, more or less continually, at this locality. The greater part of the work has been done by hydraulic mining in saprolite material on the side of a hill, and several large excavations have been made. The gold-bearing zone is several hundred feet in width, the best values being restricted to more or less parallel bands within the zone limits. The prevailing country rock is mica schist, in which occur at some points thin bands of hornblende schist. Many stringers and lenses of quartz are found within the gold-bearing zone. There are, however, within this zone more or less parallel bands in which the quartz stringers are unusually numerous and rather regularly distributed. One of these bands, known as the Reynolds, or King vein, is probably near the center of the zone. Two other auriferous bands, or zones, have been located to the southeast of the King vein, the nearer one designated as the Queen vein and the other as the Bell vein. In a cut close to the mill house another auriferous band occurs which probably lies to the southwest of the King vein. The general strike of the zone is about N. 30° E. The large excavations made in former hydraulic work are in the side of the hill contiguous to the mill house which is located on the public road near Bean Creek. The most westerly and smallest cut is close to the mill-house. A vein or zone several feet in thickness of quartz stringers intermixed with country rock was exposed at the end of this cut farthest from the building just mentioned. Owing to the presence of water in the cut no attempt was made to secure a sample for assay. To the east of the cut just mentioned two very much larger, connected cuts have been made. The westernmost of these two cuts is connected by a ditch or drain with a smaller cut higher up the hill to the northeast, which is known as the Craig cut. In Bulletin No. 4-A, of the Geological Survey of Georgia published in 1896, the results of five assays of ore samples from this gold-bearing zone are recorded. These samples were probably taken from the large cut described above as being connected by a ditch, or drain, with the Craig cut. These assays are here quoted:

Sample No. 1____0.175 oz. $3.50 gold per ton.
Sample No. 2____0.050 oz. 1.00 gold per ton.
Sample No. 3____0.125 oz. 2.50 gold per ton.
Sample No. 4____0.130 oz. 2.60 gold per ton.
Sample No. 5____0.125 oz. 2.50 gold per ton.

At the time the property was visited, some prospect work was in progress on the zone near the summit of the hill to the northeast of the cuts above described. A pit, or open cut, a few yards in length had been sunk on the Reynolds or King vein. Work had not progressed far enough to establish the average thickness and probable value of the auriferous band, or zone, at this particular point. Two samples for assay were taken, however, in this pit from a section between three and four feet in thickness composed of vein quartz with interlaminated mica schist. In No. 1, the quartz portion only of the vein was taken and in No. 2 both quartz and schist were included without discrimination. These samples on assay yielded the following results:

Sample No. 1_____$2.06 gold per ton.
Sample No. 2_____ 2.89 gold per ton.

The results of these two assays are interesting for two reasons. They show that the rejection of the interlaminated schist reduced slightly the values obtained. The fact is also brought out that the average of these assays ($2.47) is nearly identical with the average of the five assays, ($2.42) previously given, taken from the gold-bearing zone several hundred yards to the southwest.

In addition to the cut, or pit, just mentioned on the King vein, a tunnel had been driven into the hill lower down on the southeast side across the strike of the zone for about a hundred feet. This tunnel cuts the Bell and Queen auriferous bands, or zones, and it ends at a point near where it was thought it would intersect the King vein. Panning tests showing the presence of gold were made on the Bell and Queen veins as exposed in this tunnel, but no samples were taken for assay.

Along a branch on lot 10, a considerable area of placer deposits is to be noted. The greater portion of these deposits has been mined, at some localities, more than once. While narrow strips occur, here and there, in the lowlands that, for some reason, have never been worked, and while fringes of unworked gravels occur at different points about the edges of the area, it is doubtful if sufficient virgin ground remains to warrant other than very limited mining operations. Panning tests at several points in the unworked fringes of gravels just mentioned gave very satisfactory results, the gold obtained being usually rather coarse.

The milling plant at the Jones mine is situated on the public road near Bean Creek and contiguous to the large excavations in the saprolites previously described. It is equipped with a fifteen-stamp mill and a Wilfley concentrating table. The machinery is operated by water power. At the time the property was visited, development work was being carried on by Mr. A. G. Rice, of Atlanta, under an option.

Lot 11.—This lot adjoins Lot 10 on the east. The property attained some notoriety a number of years ago by reason of a placer deposit which was discovered and mined along a small branch, a tributary of the creek traversing the adjacent extensive Monroe lowlands. According to local reports this placer deposit was unusually rich for a region in which remunerative placers were not uncommon.

Stovall Property.—This property embraces a considerable portion of the extensive lowlands lying along the lower course of Bean Creek. This creek has its headwaters on the properties of the Childs and Jones mines and empties into Sautee Creek a few miles to the southeast. Much of the placer deposits along its course has been mined; but, owing to heavy overburden and slight drainage fall, considerable unworked areas are to be found towards its lower end. If several properties at this locality could be united there may be here, as in the case of the Hardeman lowlands on the Chattahoochee River, a possible field for future dredging operations. With the aid of Mr. W. I. Stovall, together with the kind assistance of Mr. A. G. Rice of the Jones mine, a test pit was sunk at a locality selected at random near a small branch emptying into Bean Creek on lot 21, 3d district. The overburden at the point selected was from five to six feet thick in addition to a gravel stratum of from two to three feet in thickness. An area of about three square yards of this gravel stratum, together with skimming from the bed rock, was washed in a small sluice box. Owing to poor facilities for constructing the sluice box and an insufficient supply of water for washing, a portion of the gold was not saved. A little less than a pennyweight was secured as a result of the test.

Lumsden Property.—This property, embracing a part of lot 44, 3d district, includes a portion of the lowland along Bean Creek lying immediately below the property last described. It is owned by Mr. J. R. Lumsden, of Sautee, Ga.

Monroe Lowlands.—Other extensive lowlands occur on lots 27 and 28, 6th district, traversed by a small creek. Considerable placer mining has been done in the past in deposits on lot 27 towards the lower end of this tract of valley land. Near a shoal in the creek a channel was blasted out in the bed rock for a number of yards for the purpose of securing steeper grade for draining and mining the auriferous deposits above on a more extensive scale than they had ever been worked previously. For some reason, these plans were never carried out. Towards the upper end of the valley, on lot 28, as far as could be judged from appearances, little mining has been carried on. Rather heavy overburden prevented the tests made at this locality from being sufficiently reliable to draw any conclusions as to the probable value of the deposits. This property is owned by Mr. W. I. Stovall, of Sautee, Ga.

HABERSHAM COUNTY

The Dahlonega belt traverses a narrow portion of the northern part of Habersham county lying between White and Towns counties on the west and Rabun county on the east. The belt is narrower in Habersham county than in White and for a distance of five or six miles northeastward along its course from the most easternly situated mines in the latter county no gold mining at all has been prosecuted. It is to be noted, however, that mica schist and gneisses, associated with hornblendic rocks, characteristic of other portions of the Dahlonega belt, occur here, and quartz veins are not wanting. The lack of mining developments on this particular portion of the belt may be due as much to chance, and the fact that the area is rather distant from any mining centres, as to the absence of valuable deposits.

PLACER DEPOSITS ALONG THE SOQUE RIVER.—A very limited amount of work has been done on placer deposits at two or three places along the Soque River and smaller tributary streams within the area of the Dahlonega belt. A little placer mining has been done in the past on the property of Mr. J. P. Woods. About seven years ago, Mr. Thomas Wilson worked out a pit about thirty feet square in the gravel deposits along the Soque River on lot 28, 11th district. He reported that very good values were obtained. Some placer mining of a more extended character was carried on a number of years ago on the property of Mrs. Lindy Wilson, lot 20, 11th district, a little to the northeast of the last mentioned locality.

Hood Mine.—This mine, on lot 22, 13th district, is situated on the crest of a high ridge at the headwaters of the Soque River. A limited amount of prospect work was done here on an auriferous quartz vein some years ago. Several small open cuts, or pits, were made and a tunnel driven into the side of the hill. Panning tests made at one or two of the test pits indicated ore of a very good quality, but until further development work is prosecuted no very satisfactory surmises can be made as to the probable size of the vein. As seen in most of the excavations it did not give very encouraging prospects for large quantities of ore. A small placer deposit occurs along a branch a short distance below the vein. Placer mining was carried on here many years ago. From gravels that had never been worked some very satisfactory panning tests were made at this locality. While some good returns might be secured here from washing on a small scale, it is not probable that there is any field for extensive placer operations.

RABUN COUNTY

The Dahlonega belt traverses the northwest portion of Rabun county passing into Macon county, N. C., at a point where the Tallulah Falls Railway intersects the State line. In Rabun, as in Habersham county, its width is much less than in White and Lumpkin counties. Mining developments along the belt in this county have been very limited. This is due in part, no doubt, to natural conditions, the topography of the region being mountainous which makes this district more difficult of access than any other section of the Dahlonega belt.

The same association of rocks rich in hornblende with

more acidic schists and gneisses is noticeable here as along
the belt in the counties to the southwest. About two hundred
yards north of the Georgia-North Carolina line, in a cut on
the Tallulah Falls Railway, in Macon county, N. C., a good
exposure of the former class of rocks is to be seen.

REAVES PROPERTY.—Some vein mining was conducted on
this property, southwest side of lot 105, 5th district, before
the Civil war, by Dr. M. F. Stephenson, an enthusiastic
student, at that time, of Georgia's mineral resources and the
author of a small book on the geology and mineralogy
of the State. The mining was conducted on an auriferous
quartz vein occurring in a hill about a fourth of a mile from
Wildcat Creek. Several tunnels were driven and shafts sunk
in the course of these early operations. About ten years ago
Mr. R. K. Reaves, of Athens, Ga., the owner of the mine, sunk
a shaft to a depth of about fifty-seven feet and drifted a short
distance to the vein. According to Mr. McGalbus, who resides
on the property, the vein as exposed was well defined and
something less than two feet in thickness. As none of the
underground works were accessible at the time of visit the
vein could not be inspected.

ROCKY BOTTOM SMITH PLACER.—Near the Reaves property
some mining operations have been carried on in the past on
placer deposits on a tract of valley land traversed by Moccasin
Creek. These operations were of a limited character and it
could not be ascertained what gold was secured.

STONESYPHER MINE.—This mine, in the northeastern part
of lot 105, 5th district, is located near the Stonesypher residence on the east side of Moccasin Creek. In addition to
some placer mining in the lowlands along the creek, prospect
work has been done in the past on several quartz veins
occurring in a ridge back of the Stonesypher dwelling. A

stamp mill was erected at the locality and operated for a short time. No work has been done here for about ten years and little can be stated concerning the character of the deposits. It is reported that some of the ore obtained in the former workings was of very good grade.

SMITH MINE.—This mine, on lots 103 and 104, is a short distance to the northeast of the Stonespyher mine and about a mile and a half west of Burton post office. Both vein and placer mining have been carried on here, and at the time of visit some prospecting on veins was in progress, under an option by Mr. J. H. Derrick, of Clayton, Ga.

Considerable placer mining was done at one time along Dick's Creek and some small branches on the property. A portion of the deposits along the creek, where the overburden is rather heavy, appears never to have been mined. Opportunity was not afforded at the time the property was visited to make any tests as to the probable value of these deposits.

The most of the vein mining has been conducted on a vein at the base of a hill toward the upper end of the old placer. An open cut was made for a distance of about a hundred yards and some shafts sunk. No exposure of the vein is to be seen at the locality, but it is reported that some very rich ore was obtained. A stamp mill was located on the property at one time. Several attempts have been made in recent years to expose the vein in these old works, but up to the time of visit none of these efforts had proven very successful. About a half a mile northeast of these old works on lot 103, Mr. J. H. Derrick, of Clayton, Ga., was doing some prospect work on a quartz vein at the time of visit. A shaft had been sunk on the vein to an approximate depth of thirty feet and a tunnel was being driven into the hillside with the aim of intersecting the vein. The vein had not been exposed in this tunnel and no attempt was made to obtain an assay sample from the shaft.

BLALOCK PROPERTY.—The Blalock property lies a few miles to the northeast of the Smith mine. A limited amount of mining has been carried on here in the past in a placer deposit along the Tallulah River. No work has been done at the locality for a number of years.

MOORE GIRL MINE.—This mine, located on lots 58 and 59, 1st district, is near Persimmon Creek and about six miles from Clayton, the county seat of Rabun county. Some years ago some open cuts, or pits, were made on, a large quartz vein close to the public road and a few hundred yards from the Moore residence. A small stamp mill, some remnants of which are still in place, was erected on the opposite side of the road on a small branch.

As exposed in the principal cut, the vein shows an ore body of solid quartz twenty feet or more in thickness. The outcrop of the vein is also prominent in the neighborhood of the cut. In the main cut, owing to limited exposure, it is difficult to determine the exact strike of the vein. A sample for assay, however, was taken from a section judged to be approximately across its trend. This sample yielded on assay $2.89 per ton. Taking into consideration the size of this vein, and the fact that these large veins are found in some cases to carry seams, or bands, that afford a much higher grade ore than the average of the whole quartz body, it would seem that this one might repay careful testing.

BARCLAY MINE.—This mine is situated near Persimmon Creek a mile or so from the Moore Girl mine and about eight miles from Clayton. Prospect work on an auriferous vein was carried on at the locality a number of years ago, and quite recently a company was formed, and it was stated, at the time of visit, that regular mining operations were soon to be inaugurated.

The mining operations above mentioned consisted principally of a tunnel driven for some distance into the side of a hill and cutting an auriferous quartz vein having a northeast-southwest trend. From this tunnel a drift was driven southwest on the strike of the vein for about thirty feet. A drift was also made for a few feet along the vein in a northeast direction, at which point a pinch occurred. Timbering prevented the vein from being inspected in the southwest drift, except at the breast. As here exposed, the vein showed a thickness of between two and three feet. Owing to the fact that a few feet of the drift at its end had not been timbered, and caving having occurred in the thoroughly decomposed wall rock, no attempt was made to secure a sample for assay. From an ore pile of a number of tons at the mouth of the tunnel representing ore taken out in driving the southwest drift above described an assay sample was taken. This yielded $129.19 per ton. Several old shafts a few yards southeast of the vein just described indicated the existence of a parallel vein, but no exposures were noticed. The tunnel previously mentioned was extended beyond the more northwestward vein with the aim of cutting the last mentioned ore body, but for some reason work was suspended before it was reached.

At the time of visit, work had commenced on the erection of a mining plant and regular mining operations were contemplated in the near future. The work to be carried on here will be regarded with much interest, as the operations if successful, will do much to encourage future mining developments on a portion of the Dahlonega belt that has received a very limited amount of attention in the past. The mine is controlled by the Barclay Mining Company, of Clayton, Ga.

LOTS 190 AND 191.—Some placer mining has been conducted

in the past along a small stream near the line between these two lots in the 2nd district on the properties of Mr. J. M. Dillard and Mr. B. C. Garland. The locality is in the Tennessee Valley about a fourth of a mile from the Georgia-North Carolina line, and is of interest as being the most northeastward point at which gold has been mined on the Dahlonega belt in Georgia. With the kind assistance of Mr. Dillard, bed rock was exposed at two or three localities and panning tests made of the auriferous gravels. A fair amount of rather coarse gold was obtained in these tests. It is not probable, however, that there is sufficient area here of unworked gravels to warrant other than very limited mining operations.

THE HIGHTOWER CREEK BELT
TOWNS COUNTY

This belt, as known at present, is a small gold belt in the northeastern part of Towns county. Commencing in the vicinity of Visage it extends northeastward to near the Georgia-North Carolina line. Mining operations have been confined to a small section of the belt in a region near Hightower Creek.

BROWN PROPERTY.—Gold is reported as having been obtained in limited quantities on the original Brown homestead situated near the headwaters of Hiawassee River. From here northeastward for several miles no prospecting seems to have been done until the Hightower Creek region is reached.

PROPERTIES NEAR HIGHTOWER CREEK.—The Newton mine, the Chastain Branch mine, the Smith mine and the Willis Creek mine are all placer deposits in the Hightower Creek region of the belt near its northeast end. The Newton and Smith mines, on lot 131, 18th district, are placer deposits along a small tributary of Hightower Creek. Some active

placer mining was carried on here a number of years ago.
The width of the deposit, where worked, was quite limited,
the stream flowing in a narrow valley. Several different
parties have worked here at different times, water for washing
having been brought by ditches from Chastain Branch, a small
stream in the vicinity of the mine. It is reported that a good
deal of the gold obtained was quite coarse, nuggets having
been found weighing as high as fifteen and twenty pennyweights. The deposit, while limited in extent, seems to have
yielded good returns, it being reported that between five and
ten thousand pennyweights of gold was secured. Along the
lower portion of the branch in the main lowlands of Hightower Creek slight drainage fall and heavy overburden have
deterred mining operations. Mr. E. E. Eller, who resides
on the property and is familiar with the history of the mine,
is of the opinion that considerable gold remains in this unworked area.

On a hill on the west side of Hightower Creek valley and
a short distance from the Eller residence, some prospect
work has been done in recent years on quartz veins. Two
or three exposures seen here in shallow excavations did not
look encouraging, the vein quartz showing very little sulphides.

The Chastain Branch mine is on lot 136, 18th district,
immediately south of the Newton mine. Placer deposits
occur here along a small branch and it is reported that a
considerable amount of gold has been secured in the course
of the mining that has been carried on.

The Willis Creek mine is on lot 102, 18th district. The
deposit is located near the head of a narrow hollow and the
portion that has been worked with profit extends over a very
limited area. It is reported that some portions of the deposits
where mined were quite rich.

On lot 94, 18th district, a limited amount of placer mining

has been done along a small stream emptying into Hightower Creek from the west.

COOSA CREEK BELT

UNION COUNTY

The Coosa Creek gold belt extends from near the head-waters of Coosa Creek in the western part of Union county, northeastward into Towns county. It passes in its course through Blairsville, the county seat, and close to Track Rock in the eastern part of the county. While a limited amount of vein mining has been carried on at several points on the belt the principal yield of gold has, so far, been derived from placer deposits along Coosa Creek. Recent prospecting has located several auriferous veins a mile or so to the northwest of the belt in the vicinity of Blairsville, and, as the topography of the country is very rugged, it has not yet been thoroughly prospected. Future developments may show that the dimensions of the Coosa Creek belt are considerably greater than as mapped in the text. The placers along Coosa Creek have yielded large amounts of gold and the mountain slopes of that region deserve careful prospecting.

PLACER DEPOSITS OF COOSA CREEK.—Placer deposits along Coosa Creek have been mined at different intervals for a long period of years. As this creek flows through a very mountainous section the extent and character of the deposits varies considerably at different points. At some localities, where the grade of the stream is not very steep, comparatively broad valley lands occur, while at other points, especially near its head, the stream course is in V-shaped defiles and the auriferous deposits are confined to very narrow alluvial strips or to the bed of the creek only. The most extensive mining operations have been conducted on lots 85, 86 and 87, 9th district, but the deposits have been mined from near the headwaters of the stream high up on a mountain side, for a distance of several miles to a point on a lower part of the stream where the deposits, apparently, ceased to be remunerative. Some of the broader alluvial borders have not been entirely worked over and at time of visit rewashing of the deposits on a small scale was being carried on at one or two localities. Whether portions of the deposit that have already been mined, together with the unworked areas, would repay hydraulic mining on an extensive scale is problematic. In Bulletin No. 4-A, of the Geological Survey of Georgia published in 1896, it is stated that at two or three points at this locality where the deposits had never been mined, test pits three feet square were sunk by the Survey and the gravels panned. Each pit yielded from ten to twenty-five cents worth of gold. The entire output of the Coosa Creek placers has been variously estimated at from a half to a million pennyweights of gold. The gold of this placer deposit seems to be unusually fine. It is stated by those who have worked in the mines that its purity is as great as .980.

Close to the headwaters of the creek, on Wellborn Mountain, prospecting has been done at different times in endeavors to locate vein deposits. About twelve years ago on lot 123, 10th district, a tunnel was run into the hillside near the top of Wellborn Mountain. Some small stringers, or seams, of quartz exposed in this tunnel are reported to be auriferous. Several hundred yards southwest of the locality just mentioned a shallow cut known as the Lunsford cut is to be seen which was made some years ago in the course of prospecting. Limited operations had recently been in progress in this vicinity at time of visit, but until the work has progressed further nothing definite can be stated concerning the results.

Work of a similar character has also been done on a hill above Owl Town Gap.

PLACER DEPOSITS NEAR CRUMLEY CREEK.—A few miles northeast of Owl Town Gap some placer deposits occur on a branch flowing into Crumley Creek. A limited amount of placer mining was carried on here a number of years ago.

LEGAL TENDER MINE.—This mine, on lot 304, 9th district, is situated near the corporate limits of the town of Blairsville. Regular mining operations were carried on at this locality eight or nine years ago by the Legal Tender Gold Mining Company. Several shafts were sunk on an auriferous quartz vein and some drifting and stoping done. A milling plant was erected near the mine and a reservoir constructed to which water was brought from a creek about a fourth of a mile distant. At the time of visit work had been suspended for some years and no opportunity was afforded to inspect the ore body. It is stated that the vein was drifted on for about seventy-five feet during the course of mining operations and that it had an average thickness, where exploited, of about three feet. The vein is said to be composed of quartz with more or less interlaminated mica schist. According to reports, the values obtained from the ore milled did not meet expectations, but it is stated by parties familiar with the mine that the values in the vein, as shown by assay, were much higher than those secured in milling.

The mine is equipped with a substantial mill house, containing a Huntington mill and several Frue vanners. The property is controlled by Mr. John S. Martin, of Chattanooga, Tenn.

BUTT PROSPECT.—On lot 272, 9th district, about a mile to the northeast of the Legal Tender mine, some prospect work was done on an auriferous quartz vein several years ago by

the late Mr. James Butt, of Blairsville. A shaft was sunk
on the vein a few yards from a small branch flowing through
the property. At time of visit, owing to caving and filling in,
the vein was not accessible in this shaft, but according to Mr.
C. E. Butt, of Blairsville, it is a well defined one. A sample
for assay was taken from an ore pile at the mouth of the
shaft. This sample yielded on assay $7.85 per ton. Several
other shafts, or pits, have been sunk on this lot, but the prospect described is considered the most promising. Mrs. Butt
has in her possession a small button made from gold secured
during her husbands' lifetime from mortaring and panning
some of the ore from this property. · The prospect is favorably
regarded by those familiar with the auriferous deposits of
the Coosa Creek belt.

ROGERS AND STEPHENS PROPERTIES.—About a couple of miles
northeast of the last described property, some prospect work
has been done on quartz veins on the property of Mr. Cicero
Rogers and also on the adjoining property of Mr. Thomas
Stephens. The work at both localities was done several years
ago and nothing very satisfactory can be stated as to the
character of the veins. Some panning tests of ore picked
up at the old works gave unsatisfactory results, but, owing
to the lack of a guide, these tests were made at random
without information concerning the deposits.

KILLIAN PROPERTY.—On the property of Mr. W. M. Killian,
lot 214, 17th district, some prospect work has been done at
different times on several quartz veins. Insufficient exposure,
however, at the time of visit prevented any definite conclusions
being reached concerning the probable value of these veins.

ROSS PROPERTY.—On the summit of a high ridge on the
opposite side of the public road from the Killian residence
a shaft was sunk in 1904 by Mr. W. M. Killian on a large

FIG. 1.—BLAST FURNACE, SEMINOLE GOLD AND COPPER MINE, LINCOLN COUNTY, GEORGIA

FIG. 2.—VIEW OF A PORTION OF ROASTING AND SMELTING FURNACE, SEMINOLE GOLD AND COPPER MINE, LINCOLN COUNTY, GEORGIA

quartz vein outcropping strongly along the crest of the ridge. Several panning tests made from the ore yielded unsatisfactory results.

LOT 213.—Several years ago some prospect work was done by Messrs. W. M. Killian and S. C. Rhodes on a small quartz vein, occurring on lot 213, 17th district, about two miles to the northeast of the Ross property. No opportunity was afforded to inspect the vein in place and very little could be ascertained concerning the character of the deposit.

LOT 184.—On the side of a high ridge a half or three-fourths of a mile from the residence of J. P. Bowling, on lot 184, 17th district, an auriferous quartz vein occurs that was prospected about 1904 by Mr. W. M. Killian. This vein has a northeast-southwest strike and dips into the ridge at a steep angle. A shaft was sunk on the incline of the vein to a depth of about seventy feet. This shaft, at the time of visit, could only be explored for a short distance. The vein had a thickness at the mouth of the shaft of two and a half or three feet, but was thinner lower down. Some specimens of ore from the vein near the surface were mortared and panned. A small amount of rather fine gold was secured.

WELLBORN-ROBINSON PROPERTY.—About two miles northeast of Blairsville, and approximately the same distance northwest of the regular trend of the Coosa Creek belt, Messrs. E. C. Wellborn and Fletch Robinson had recently conducted some prospect work on an auriferous quartz vein. The vein, as exposed in either side of a tunnel that had been driven across its strike, was rather ill defined and showed an average thickness of eight or ten inches. An assay sample taken from this exposure yielded $2.06 per ton.

TOWNS COUNTY

The Coosa Creek gold belt has an extent in Towns county of only a few miles in the portion adjoining Union county.

MALDEN PROSPECT.—On lot 99, 17th district, about a mile west of the town of Young Harris, some prospect work was done on an auriferous quartz vein several years ago. The results of an assay of a sample from this vein are given in Bulletin No. 4-A, of the Geological Survey of Georgia, published in 1896. It is stated that where sampled the vein had an average thickness of about six inches. The assay showed a value of $2.50 per ton.

THE GUM LOG BELT

UNION COUNTY

The southwestern end of the Gum Log gold belt occupies a small extent of territory in the northeastern corner of Union county. As in the case of the Coosa Creek belt, the topography of the region traversed is extremely rugged and it has never been thoroughly prospected. Sufficient work has been done, however, at one or two localities to demonstrate the occurrence of some ore of a good grade.

GUM LOG MINE.—The Gum Log mine is situated at the end of a low ridge on lot 59, 9th district. The location is about nine miles northeast of Blairsville. Mining on a small scale has been carried on here at intervals for half a century. Several veins occur in the ridge but, owing to very limited exposures, no definite conclusions could be reached concerning their probable size and character. As seen on the end of the ridge where old surface work gave some exposure, the ore bodies appeared to be in the nature of thin interlaminated veins or stringers affording larger bunches of quartz at some points. It is stated, however, that, in some of the old works, veins several feet in thickness were mined. A number of old shafts and cuts are to be seen on the ridge and along a branch at the base of the ridge a small area of placer deposit has been mined. Some pieces of float ore picked up about the works on the ridge and mortared and panned yielded good results. At the time of visit, Messrs. E. C. Wellborn and Fletch Robinson were installing a stamp mill and preparing to clean out several of the old shafts. Some ore from the bottom of one of the shafts, furnished by these gentlemen, when mortared and panned, yielded extremely satisfactory results.

Much of the vein quartz at this locality is laminated and of a granular texture and easily crushed, its appearance suggesting that the veins had been subjected to considerable shearing stress at, or subsequent to, the time of their formation. The deposit here is an interesting one and well deserves careful and thorough investigation.

Brown Mine.—This mine, on lot 19, 9th district, is northeast of the Gum Log mine. Considerable prospect work, consisting of shafts and tunnels, was done here a number of years ago. An examination was made of the old works, but no satisfactory conclusions were reached as to the character and probable value of the deposits.

Hunt, or Princeton, Mine.—This mine, on lot 55, 9th district, is a short distance to the northeast of the Brown mine and close to the Union-Towns county line. Underground mining on a limited scale has been carried on here at different times. About 1900 the Princeton Mining Company sank a shaft to a depth of about eighty-five feet on an auriferous quartz vein occurring at this locality. A short distance to the west of this shaft an open cut has been made and from the bottom of this a shaft was sunk in 1906 to a depth of forty feet.

As exposed in the open cut the vein shows a thickness of about three feet and is composed of quartz with interlaminated wall rock. The general strike of the vein is nearly east and west. Neither of the two shafts were accessible at the time of visit. Mr. Frazier Gillman, who conducted the last mining done at the locality, exhibited some specimens of ore showing free gold that he stated had been taken from the bottom of the forty-foot shaft. A small test stamp mill, boiler and hoisting engine are located near the shaft sunk by the Princeton Mining Company.

WELLBORN HILL MINE.—This mine, on lot 18, 9th district, is a short distance from the last described property and close to Gum Log Creek. Considerable underground mining was carried on here in the eighties. The principal work was done near the summit of a low ridge. A shaft was sunk to a depth of over a hundred feet and it is stated that several hundred feet of drifting was done. According to reports, portions of the vein yielded a high grade ore. At the base of the hill, several hundred yards northeast of the main shaft ,a tunnel was driven into the hill on the vein for a distance of over a hundred feet. No mining operations of any importance have been carried on at the locality for a number of years. In Bulletin No. 4-A, of the Geological Survey of Georgia, published in 1896, it is stated that the vein, examined by the Survey in a shaft seventy-five feet west of the main shaft, showed a thickness of from eighteen to twenty-four inches and consisted of light colored, somewhat granular quartz with numerous iron stained cavities.

TOWNS COUNTY

The Gum Log gold belt has an extent of a few miles in Towns county in the extreme northwest corner. The deposits are similar in character to those of the Union county portion of the belt.

LOT 54.—Some prospect work was done on lot 54, 19th district, close to the Towns-Union county line, several years ago by the Blue Ridge Gold Mining Company. A tunnel was driven into the side of a hill for about a hundred feet and a quartz vein exploited. The property was examined, but, as no work had been in progress for some time, nothing definite could be ascertained concerning the character of the vein.

GREATER PITTSBURG MINE.—This mine, on lot 38, is close to the last described property. The Greater Pittsburg Gold Mining Company conducted mining operations a few years ago on a quartz vein at this locality. A shaft was sunk and some mining machinery installed. No work was in progress at time of visit and nothing definite was ascertained concerning the character of the deposit.

NANCY BROWN MINE.—The Nancy Brown mine is on lot 34, 17th district, a short distance northeast of the Greater Pittsburg mine. Different parties have conducted underground mining here at intervals since 1874. Mr. Frazier Gillman, the present owner of the mine, has in late years carried on some test work, but no extensive mining operations have been in progress for some time. An auriferous quartz vein has been exploited at a number of points along its strike, but at time of visit the underground works were inaccessible and few surface exposures were noticed. The main, or Gillman, shaft is about a hundred and twenty-five feet deep and is a two compartment shaft, one compartment being equipped with steps for the safety of the miners. At the time of publication of Bulletin No. 4-A, of the Geological Survey of Georgia, published in 1896, the vein was exposed to observation in a number of shafts and pits. The following description of the ore body, by State Geologist S. W. McCallie, is therefore quoted from that publication:

"There appears here a more or less continuous gold-bearing vein extending diagonally across the northwest corner of the lot, parallel with the strike of the mica schist. It varies greatly in size and in the character of the ore at the various exposures. At the opening furthest north, near the small stream which flows from the east across the lot, the vein is from three to five feet wide and is made up of numerous thin layers of quartz interlaminated with mica schist, while at the opening furthest to the south, it consists of a compact, milk-colored quartz from eighteen to twenty inches in thickness. The dip of the ore body corresponds to the country rock, but at some of the exposures there occurs an unconformity between the overhanging and foot walls that is evidently due to local faulting.

"Williams and Pruett began mining operations on this property in 1874 and worked the mine, though not continuously, for about four years. During this time several tons of ore were taken out and hauled two miles to a stamp mill which had been erected by Perry Ellis near Welsh. This ore is said to have milled, on an average, $18.00 per ton and was taken from the north exposure of the vein where a tunnel, several yards in length, was driven into the hill along the vein at water level. Below this level, the vein is said to continue at its usual width and richness, but, owing to inadequate drainage and the falling in of the tunnel walls, it was found impracticable with the means then at hand to prosecute the work further."

Six samples were taken by Prof. McCallie from pits on the Nancy Brown vein and assayed in the Survey laboratory with the following results:

1. Ore sample, pit No. 1, .10 oz. ($2.00) of gold per ton.
2. Ore sample, pit No. 5, trace of gold
3. Ore sample, pit No. 7, .05 oz. ($1.00) of gold per ton.
4. Ore sample, pit No. 7, .125 oz. ($2.50) of gold per ton.
5. Ore sample, pit No. 7, .125 oz. ($2.50) of gold per ton.
6. Ore sample, pit No. 8, .075 oz. ($1.50) of gold per ton.

OLD FIELD MINE.—On lot 3, immediately. to the north of the Nancy Brown mine, considerable prospect work has been done at different times at a locality known locally as the Old Field. Gold was discovered here many years ago in float ore and in residual surface material. A number of shafts and pits have been sunk and also some tunnels driven in efforts to locate ore bodies that could be worked with profit. It is stated that a vein was found, but the prospects that were located seem never to have warranted permanent operations. At the time of visit, owing to caving and filling in, no exposures were noticed in the old works.

LOTS 1 AND 2.—On both of these lots, in the 17th district, a little mining for gold has been carried on in the past. In the southeastern portion of lot 1 some prospect shafts, or pits, were sunk on a vein designated as the Horse vein. As the work was done years ago, nothing definite can be stated concerning the character of the deposit. On lot 2, in addition to some vein prospecting, a placer deposit occurring along a small stream was mined before the Civil war.

MURDOCK VEIN.—The Murdock vein is located on lots 32 and 42, 17th district. Gold was discovered here shortly before the Civil war and mining operations inaugurated soon afterwards. Work was suspended during the period of the war just mentioned, but was renewed at a later period. During the early operations an arrastre was used for milling the ore. This was subsequently replaced by a ten-stamp mill. In 1884 the property came into the hands of the Hiawassee Gold Mining Company, who are reported as the present owners. No work has been done at the locality for a number of years and an examination of the property afforded little data from which to draw conclusions concerning the character of the ore body. Old open cut works are to be seen along the strike of the vein for a hundred yards or more and it is stated that some ore of a high grade was secured during the course of mining operations. In Bulletin No. 4-A, of the Geological Survey of Georgia, published in 1896, the following description of this vein is given:

"The Murdock vein is what is known as a true fissure vein; it cuts the mica schist, or country rock, at nearly right angles. The stirke is almost due northwest while it dips at a high angle to the northeast. The vein varies in thickness from six inches to two feet, and it may be traced with a considerable degree of certainty for a quarter of a mile. In places it is much fissured and broken, as if it had been subjected to great crushing force. The quartz is more or less iron stained and it frequently contains cavities in which free gold may be seen."

LOT 67.—Some mining operations have been conducted on an auriferous quartz vein on lot 67, 17th district, about a mile from Welsh. A tunnel was driven into a hill for a distance of about a hundred feet and some shafts sunk. This work was done a number of years ago and nothing very definite can be stated concerning the character of the vein.

ISOLATED LOCALITIES

LINCOLN COUNTY

SEMINOLE MINE.—The Seminole mine is situated in the western part of the county immediately contiguous to the

Wilkes county line and about two and a half miles east of Metasville in the last named county. A series of auriferous, cupriferous deposits occur here on a line of intense shearing, and mining operations have been conducted at the locality at intervals and by different parties through a long period of years. Gold was discovered prior to the Civil war along a small stream near the mine and vein mining was begun soon after this discovery. Until recent years the mine was known as the Magruder mine after one of the early operators. In the seventies the mine was owned and worked by a Mr. Jackson, of Augusta, Ga., and during his management copper and lead, in addition to gold, became to be regarded as valuable products of the deposits. Later, the Seminole Mining Company operated the mine erecting under the supervision of Mr. Carl Hendrich, the general manager, a combined roasting and smelting furnace and also a small blast furnace. Copper, gold and silver values are secured in the mattes.

As previously stated, the ore bodies occur along a shear zone and consist principally of three roughly parallel veins designated as the Wardlaw, the Findley and the Magruder veins. The general strike, both of the veins and the country rock, is northeast and southwest. The country rock in the vicinity of the mine is of an unusual type and can be seen in a fairly fresh condition in the ravine of a small branch a hundred yards or more to the northwest of the Magruder vein, the northwesternmost of the three veins above mentioned. This rock contains large numbers of spherical quartz anhedra, some of them showing imperfectly developed pyramidal crystal faces, and the residual material of the rock can be identified over large areas by the presence of these quartzes strewn unaltered over the surface. By this means its saprolite can be traced for a considerable distance to the west of the ravine above mentioned. Southeast of the mine for some distance the surface residual material affords little clue to the character of the underlying rock, but something like an eighth or quarter of a mile in that direction the quartzes spoken of are to be found in the soil over an extensive area. Microscopic study of a specimen of this prevailing rock, from an exposure in the ravine previously mentioned at a point something over a hundred yards northwest of the Magruder vein, shows an igneous rock possessing the structure of a granite porphyry. Numerous quartz anhedra together with phenocrysts of albite and flakes of biotite, are scattered through a fine grained ground-mass of quartz and plagioclase feldspar in intimate intergrowth. Some of the ground-mass has a rather coarse spherulitic structure, such as is sometimes seen in small dikes or in larger bodies of certain igneous rocks near their peripheries. The rock is porphyritic, but shows, both in the hand specimen and under the microscope, distinct evidences of pressure. A chemical analysis indicates over six per cent. of soda and a little less than one per cent. of lime and still smaller amounts of potash. A short distance further down the ravine at a point about thirty yards from the extension of the Magruder vein a portion of the same body of rock is to be seen, but here it is much more schistose in structure and the phenocrysts of feldspar and the quartz anhedra have largely disappeared. Close inspection, however, will reveal in the hand specimen a few of the characteristic quartzes. Considerable white mica is present as a new mineral, many of the areas of the feldspar phenocrysts being nearly filled with this material. Further to the southeast, within the zone of the ore bearing veins, the rock is a very fine grained, highly friable, light colored schist which is seen under the microscope to be composed principally of quartz and white mica. Some biotite is still present and remnants of the quartz anhedra are also discernable. The silica content has increased from sixty-eight per cent. in the specimen first described to over seventy-seven per cent. in this light colored schist. All gradations may be traced at points in the underground workings between this quartzose schist and the vein quartz though, at some localities, solid well defined bodies of vein quartz are found. The veins appear to owe their origin to alteration and silicification of the quartz-albite porphyry along several lines in a shear zone accompanied with fissuring and the deposition of vein quartz in cavities in addition to the metasomatic action. Conditions for the circulation of ore bearing solutions and the deposition of metallic sulphides were unusually favorable, as at no other gold mine in the State has such a variety of metallic sulphides been found in quantity in a single deposit.

Associated with the ore bodies and intersecting their strike, are basic dikes representing a later period of dynamic disturbance. These dikes are schistose and of a dark color and present a striking contrast to the light colored schist. A section of a specimen of one of them taken from the dump near the main shaft of the mine shows under the microscope an igneous rock composed principally of hornblende and basic feldspar with some quartz. Hornblende of the common green variety in fibres, and ragged elongated prisms, makes up over half of the rock. Some of the prisms of this mineral on cross section show fairly well developed cleavage. Much of the quartz has the appearance of being secondary. These dikes probably vary considerably in mineral composition at different points, as Dr. Thos. L. Watson, in a paper on the Seminole copper deposit, says in describing them:

"**** They are very dark and are composed chiefly of

1. Watson, Thos. L., The Seminole Copper Deposits of Georgia, Bulletin No. 225, U. S. Geol. Surv., 1903, pp. 182-186.

altered feldspar, some biotite, hornblende, and much quartz, which in most cases at least is probably secondary.''[1]

The dikes do not seem to have had any special influence on the deposition of the gold and the copper minerals, though, as pointed out by Watson in the paper cited, the dynamic disturbances represented by their intrusion and subsequent shearing probably aided in mineralization. As an evidence of an unusual amount of mineralization in the shear zone at this locality, small crystals of pyrite are of common occurrence both in the light colored schist and in the dikes and also in the less sheared quartz-albite porphyry exposed in the ravine to the northwest of the mine.

As previously stated, three principal veins, or ore bodies, are recognized at the Seminole mine and they have been desig- nated as the Magruder, the Findley and the Wardlaw. The Magruder is the most northwesterly located. Lenses of silicified and mineralized schist occurring between the veins may also be considered as ore bodies. The general strike of veins is northeast and southwest and the dip is nearly vertical or quite steep. The thickness of the veins varies at different points and, owing, at many localities, to lack of definite walls, familiarity with the values are necessary in order to decide what should be considered vein, or ore bearing material. Some measurements of the Wardlaw vein at different points in the hundred and forty-five foot level indicated a thickness of eight or nine feet. In addition to gold, pyrite, chalcopyrite, galena and sphalerite are found in the ore bodies. Silver is present, probably occurring as argentiferous galena, and it is stated by reliable parties that in former mining operations masses of native copper weighing as high as sixty pounds were obtained from near water level. Pyro- morphite has been noticed at a few points in the veins, occur- ring as small crystals in seams or crevices in the quartz. The sulphides are distributed rather irregularly through the veins.

At some points in the Wardlaw, and also in the Findley vein, bands have been encountered very rich in chalcopyrite and galena. According to reports these bands are in some cases as much as two feet in thickness.

At the time of visit the main shaft had penetrated to a depth of over two hundred feet and drifts had been driven at ninety, a hundred and twenty-five, a hundred and forty-five and a hundred and eighty-five foot levels. The mining for gold carried on before the Civil war was conducted chiefly on the Magruder vein. In addition to some shafts and open cut work, a tunnel was run for a distance of about seven hundred feet from a point near a branch to the Magruder vein striking a shaft, known as the Magruder shaft, at about sixty feet from the surface. For what purpose this extensive tunnel was driven is not apparent. A portion of it now constitutes the ninety-foot level of the mine. The amount of gold obtained from these early operations can not now be ascertained. Under Jackson's management, in the seventies, a shaft was sunk to a depth of a hundred feet near the Magruder vein with which it was connected by a thirty-foot cross cut. Considerable drifting and stoping was also done on the Wardlaw vein. Jackson also did considerable surface work for gold on various portions of the deposit. Records of the gold obtained under this management are not available, but Mr. John McHale, a reliable employee of the mine, remembers that at one period of the operations a month's work secured five hundred dollars worth of gold.

When the mine was visited in the spring of 1908, no mining was in progress and the underground works could not be entered. At a visit to the locality several years previous to that time, the various drifts were examined. Little or no opportunity was afforded to inspect the Magruder vein, the most of the recent drifting, amounting in the aggregate to a number of hundred feet, being on the Wardlaw and Findley veins. From the hundred and forty-five foot level on the Wardlaw vein a cross cut two hundred and eighty-five feet long had been driven to intersect the Magruder vein and a drift had been run for about seventy feet from this cross cut on what was thought to represent that vein. Owing to timbering, however, an examination proved unsatisfactory. As regular mining operations were not in progress when the underground works were visited systematic sampling was not undertaken. One sample for assay was taken from a section of the Wardlaw vein in the hundred and twenty-five foot level. The sample was taken from an exposure of the vein for about seventy-five feet along its strike between a dike which intersects it approxiamtely fifty feet northwest of the main shaft and an old stope about twenty-five feet to the northeast of the shaft. This sample, assayed at the N. P. Pratt laboratory, Atlanta, yielded .15 oz. ($3.00) of gold per ton and 2.80 oz. ($1.40) of silver per ton. By an oversight, the copper values were not determined.

The milling plant of the Seminole mine is situated a short distance to the east of the main shaft. (Plate VII, Fig. 1.) The ore, after being crushed, is passed through roller mills and the various valuable products are secured as concentrates with Bartlett tables. A Century jig is used to separate the galena concentrates. A combined roasting and smelting furnace of about fifty by fourteen feet (outside dimensions) and a small blast furnace, with which the mattes are produced, are situated convenient to the mill. (Plate VIII, Fig's. 1 and 2.)

SALES MINE.—The Sales mine, or the Sales and Lamar mine, as it is sometimes called, is located in the northwestern part of the county near Fishing Creek and about an eighth of a mile from the Seedtick road from Lincolnton to White Plains.

Gold was discovered along a branch on which the mine is located before the Civil war and placer mining was undertaken and in addition a company, composed of the owners of the property, erected a stamp mill and did some milling. No information is available as to the amount of gold obtained from these early workings. About 1879 the mine was worked by a company who paid the owners a royalty. After their lease expired another company worked for a short time and was succeeded by the Sales and Lamar Mining Company with J. D. MacNeal, of Hamilton, Ohio, as president, and B. M. Hall, recently of the hydrographic division of the United States Geological Survey, superintendent of the mine. This company took out ore from a vein designated as the "Mother Lode" for a distance of approximately thirty feet along its strike and to a depth of about the same number of feet. The vein varied in thickness from one to two or more feet, but the thinner portions afforded the better grade of ore. According to Mr. Hall about $100.00 per ton was obtained from these richer portions of the vein. In addition to this work an area of probably two hundred yards in length by approximately forty in width was washed out with a hydraulic giant to a depth of from forty to sixty feet. In this area small stringers of quartz were found which afforded thin chimneys or chutes of rich ore. The hydraulic mining did not pay as well as was expected, but Mr. Hall attributes this to the fact that many of the chimneys or chutes had been worked out to considerable depths before the Sales and Lamar Mining Company began operations. This company erected a twenty-stamp mill of four hundred and fifty-pound stamps and milled the material taken out as free milling ore. When a rich chute was struck the ore was first beaten by hand in mortars and after all the gold was secured that could be obtained by this method the crushed rock was fed to the mill.

Since the Sales and Lamar Mining Company suspended operations the mine has been worked several times in a small way by different parties.

WARD PROPERTY.—The Ward property adjoins the Sales mine. About 1880 Mr. B. R. Ward sank several shafts on the supposed continuation of the veins of the Sales mine. The ore taken out was worked in sluice boxes and some milling was also carried on. It is reported that from very limited operations about six hundred dollars worth of gold was secured.

WILKES COUNTY

STONE RIDGE MINE.—This mine is situated about six miles southwest of Washington, the county seat, and is near a road connecting the Greensboro and Skull Shoals highways. Gold was discovered here in 1879 and since that time a limited amount of mining has been conducted at intervals by different parties on an auriferous quartz vein occurring along the crest of a ridge. In 1881 Mr. C. E. Smith, of Washington, Ga., carried on mining operations for a short period on a portion of the ridge known as the Marlow property. This gentleman exhibited United States mint returns for gold he obtained showing an aggregate of $1,658.15. Some mining has also been conducted on the adjacent Smith and Crosby properties.

Considerable open cut work was done on the vein; shafts were also sunk and some drifts driven. At the time of visit no exposures of the ore body were noticed in the old works. It is stated that the trend of the vein is along the crest of the ridge, the strike being about N. 15° E. Prospectors have reported the occurrence of gold along the ridge for several miles in either direction from the Stony Ridge mine.

G. W. BOOKER PROPERTY.—A little prospect work for gold has been done on the Booker property, which is situated on the French Mills public road ten miles northeast of Washington. About a fourth of a mile from the Booker residence, from some earth taken from a small excavation near a branch, a little rather coarse gold was panned when the property was visited. On the opposite side of the stream, and nearer the dwelling house, a little fine gold was panned from saprolite material in a small prospect pit. No quartz veins, other than some extremely small stringers, were noticed at either locality.

JASPER WOLF PROPERTY.—This property is situated on the French Mills public road ten and a half miles northeast of Washington. Some gold has been found here in a small stream about an eighth of a mile east of the dwelling house. A little rather fine gold was secured in a panning test at the locality at time of visit. It is stated that gold has also been obtained at a point some distance from the branch just mentioned. An old prospect shaft is to be seen at the latter locality.

KENDALL MINE.—The Kendall mine is situated on the French Mill public road eleven miles northeast of Washington. Some mining operations were carried on here in 1881 by Mr. C. E. Smith, of Washington, Ga. A shaft was sunk to a depth of seventy feet and forty or fifty feet of drifting was done. A stamp mill was also erected and about twenty-five tons of ore milled. According to Mr. Smith, $415.00 worth of gold was secured. No exposures of the vein were noticed at the surface when the property was examined and the old shaft was inaccessible.

MERIWETHER COUNTY

WILKES MINE.—The Wilkes mine is situated on the west half of lot 44, 11th district, in the northwest corner of Meriwether county. The locality is about two and a half miles southeast of Grantville, a station on the Atlanta and West Point Railway.

Gold was discovered along a branch on lot 21 about a half of a mile to the north of this mine before the Civil war. Prior to 1904 the Wilkes mine was operated at intervals by a number of different parties. Mr. John Cross, whose work extended through a period of ten years, is reported to have obtained between eighty and a hundred thousand dollars worth of gold. Later, the Wilkes Gold Mining Company, of Massachusetts, the present owners, operated the mine for about four years. The underground works have attained a depth of approximately a hundred and thirty feet and it is stated that about two hundred feet of drifting has been done. The principal underground works were inaccessible at the time of visit and no very definite information was obtained concerning the character of the deposit. It is reported that some ore of a high grade was obtained when the mine was operated.

Limited exposures indicate that the country rock in the vicinity of the mine is a contorted biotite gneiss, at some localities highly garnetiferous. At a number of points, in old prospect pits and tunnels, very coarse, highly feldspathic layers, representing possibly in some cases interlaminated pegmatite dikes, were noticed. According to those who have mined at this locality gold is frequently found intimately associated with these feldspathic portions of the rock. In the surrounding region bands, or small areas, of reddish brown saprolites were observed that may have resulted from the weathering of hornblende dikes, but, judging from the amount of quartz present in the residual material, they more probably represent portions of the gneiss that contained unusually large percentages of biotite.

The milling plant of the Wilkes mine, situated on a small branch several hundred yards from the mine, is equipped with two Sturtevant roller mills, two Kincaid mills, concentrating tables, an amalgamating tube, settling tank and other machinery.

LONE OAK MINE.—The Lone Oak mine on lot 43, which is owned by Mr. T. M. Zellars, of Grantville, Ga., adjoins the Wilkes mine. Some mining operations were conducted several years ago at this locality by the Lone Oak Mining Company. A shaft was sunk to a depth of a hundred feet and a drift was driven for a length of about a hundred feet to an inclined shaft. No exposures of the deposit were noticed. It is stated that the vein averaged about two feet in the drift mentioned. Milling machinery was installed and some milling done. According to reports the results of the milling operations were not satisfactory, though it is stated that good values were indicated by assay tests.

POST PROPERTY.—This property, on lot 21, 11th district, is situated a short distance to the north of the Wilkes mine. A small placer deposit occurring along a branch was mined here many years ago. In more recent years some vein mining was done on a hillside contiguous to the placer. Some shafts are still to be seen at the locality, but nothing definite can be stated concerning the character of the deposits. A small stamp mill was erected on the branch when the mining was done. The property is owned by Mr. W. A. Post, of Grantville, Ga.

COWETA COUNTY

BINGHAM PROPERTY.—This property is in the southern part of Coweta county close to the Meriwether county line and about three miles northeast of Luthersville, in the last named county. A placer deposit along a small stream was worked at this locality many years ago. The mill pond of Mr. T. N.

Hurst's grist mill covers a portion of the old workings. Traces of other old works are to be seen along the branch at a point several hundred yards below the mill.

CLARKE AND HILL PROPERTIES.—These two properties are in a sharp bend of the Chattahoochee River at a point about thirteen miles northwest of Newnan, which is situated on the Atlanta and West Point Railway. The two properties adjoin, and an auriferous gravel deposit occurs at the locality, which, according to Mr. L. L. Hill, of Newnan, has an extent of something like two hundred acres. A few test pits have been sunk at different points, and as seen in these the gravels showed a thickness of several feet with an overburden of several yards in thickness. On the Hill property some thinner gravel beds with lighter overburden were noticed.

Much of the area of these gravels is considerably above the present level of the Chattahoochee River, and, owing to the heavy overburden and very limited exposures, systematic tests as to the gold contents could not be undertaken. A little fine gold was panned at one or two points on the Hill property. Mr. L. L. Hill reports that at some points in the deposit as high as a hundred and fifty particles of gold have been obtained at a panning.

On the Clarke property a little prospect work has been done on quartz veins. From some ore at the mouth of a shaft that was sunk in the course of this work a sample was taken for assay. This sample yielded $2.48 per ton.

HARDAGREE PROPERTY.—A little prospect work was done on this property at the time the Clarke property was prospected. The work referred to was done by Mr. Clarke and it is reported that some gold was secured. The locality is close to the Hollingsworth Ferry road and about three miles below the Clarke property.

HENRY COUNTY

The occurrence of gold has been reported at several localities in Henry county. On the Walker property, lots 68 and 69, 7th district, an old excavation of considerable size is to be seen that, according to reports, was made in work done for gold before the Civil war. From a branch near by a little fine gold was panned at the time of visit. The locality is about three miles north of McDonough, the county seat. Gold is also reported as having been found on the Harper property, six miles northheast of McDonough, in the vicinity of Hampton and at one or two other localities.

NEWTON COUNTY

Gold has been obtained in commercial quantities in Newton county at one locality. On the Middlebrooks property, seven miles southwest of Covington, the county seat, a small placer deposit occurs which has been worked for gold at several different times, some mining having been carried on before the Civil war. The deposit, which is quite limited in area, occurs along a small branch and would not probably repay mining operations other than of a very limited character.

CAMPBELL COUNTY

Gold has been found in small placer deposits along branches in a section of the western part of the county near the Chattahoochee River. On the Camp property, lot 1, 8th district, auriferous gravels occur along a small stream tributary to Pea Creek. The locality is about eleven miles west of Fairburn, the county seat, and a half of a mile from the Chattahoochee River. In some panning tests made at the time of visit a number of particles of rather fine gold were obtained and it is possible that this deposit might repay mining operations on a limited scale.

DOUGLAS COUNTY

Roach Mine.—A limited amount of placer mining has been done on lot 2, 2nd district. The locality is near the Smith Ferry road and about nine miles south of Douglasville, the county seat. A little placer mining has been conducted at different times along a small branch. The portion of the stream's course where the mining was done is in a rather narrow ravine and the extent of the deposit is probably quite limited.

Bagget Property.—This property is located about a mile from Winston, a station on the Atlanta and Birmingham Division of the Southern Railway. Mr. J. P. Boatright, of Bremen, Ga., prospected a quartz vein at this locality several years ago. It is stated that some gold was obtained, but, owing to lack of exposures at the time of visit, nothing could be ascertained concerning the character of the vein.

Carnes Property.—This property is about a mile north of Winston. Some prospecting was done here on a quartz vein a number of years ago. It was not learned what success attended the work.

WALTON COUNTY

Auriferous placer deposits of limited extent have been located and mined in two sections in Walton county. Two or three deposits have been worked at a locality about eight miles north of Monroe, the county seat, and some placer mining has also been carried on at a locality about the same distance south of Monroe.

THOMSON PROPERTY.—An auriferous gravel deposit occurs along a small branch on this property which is about eight miles north of Monroe and close to the Gainesville public road. The branch, along which the deposit occurs is tributary to the Appalachee River and the mining was done at a point about a half of a mile from its mouth. Some work was done before the Civil war and in 1900 and 1901 a little mining was carried on at the locality. A small amount of gold was obtained in panning tests made at the time of visit.

SMITH MINE.—Some placer mining was done years ago by Mr. Wm. Smith at a locality about five miles north of Monroe and close to the Vineyard Creek public road. It could not be ascertained what gold was secured.

MALCOLME MINE.—A small placer deposit occurs on the Malcolme property on the Monroe and Madison public roads about seven miles south of Monroe. From gravels along the forks of a branch on the property a little fine gold was panned at the time of visit.

CONNER MINE.—A small placer deposit occurring on the Knox, formerly Conner, property was mined a number of years ago. The locality is about ten miles south of Monroe and on the road from Ebenezer church to Social Circle.

TALIAFERRO COUNTY

Some prospect work for gold was done many years ago at one or two localities in Taliaferro county. On the Lucas property, about four miles northwest of Crawfordville, some old works, now nearly obliterated, are to be seen at a point about a hundred yards from the Georgia Railroad. According to accounts, gold has been obtained at two or three localities in the county, but nothing definite could be learned about the probability of its occurrence in commercial quantities.

GREENE COUNTY

Gold has been found in commercial quantities at a locality in the eastern part of this county close to the Taliaferro-Greene county line and about six miles northeast of Union Point, a town on the Georgia Railroad. A low ridge of light colored, resistant rock, outcropping strongly for several hundred yards and impregnated with small crystals of pyrite has been rather extensively prospected. According to reports some mining operations were carried on here as early as 1852, a Chile mill having been in use on the property. In 1875 Mr. Wm. Smith carried on limited mining operations, washing gold from residual material at a small branch that flows near the base of the ridge. It is stated by reliable parties that he obtained from nine to fifteen dollars worth of gold per week. Panning tests made at the time of visit from residual material near this branch secured small amounts of very fine gold. A few years ago a mining company exploited the deposit systematically, driving a tunnel into the ridge for several hundred feet and running several cross drifts. A cyanide plant was also erected. Work had been suspended previous to the time of visit and it could not be ascertained what success attended these operations.

The ridge in which the auriferous deposit occurs has been locally termed quartzite, but microscopic investigations indicate that it may represent a light colored igneous rock of a granite character that has been sheared and more or less silicified and impregnated with auriferous pyrite. In thin section, this rock shows a rather even, fine grained interlocking mass of quartz and feldspar with accessory sphene in grains and irregular masses and occasional shreds of white mica. Crystals of iron pyrite are also present in small amounts. The quartz is in excess of the feldspar and much of it shows marked undulous extinction and other evidences of having been subjected to shearing and pressure. Alkalic feldspar, principally microcline, occurs in irregular shaped grains and also show evidences of shearing stress. While some of both the quartz and feldspar is in the form of rather rounded grains, no rims of secondary silica were noticed about any of the quartzes and the feldspar is unusually fresh and free from secondary minerals. Only a few patches of kaolin were noticed among the grains. Occasional areas, roughly spherical in outline, and composed of both quartz and feldspar, occur that suggest original clastic grains. The study of a number of specimens would probably determine with certainty the question of origin.

The main tunnel driven by the mining company above mentioned passes for several hundred feet through the rock that has been described and for a short distance beyond into a gneiss or mica schist formation. A sample for assay was taken from a section of this tunnel from the mouth to the mica schist or gneiss just mentioned and also from a short drift leading off to the left. This sample yielded on assay $1.24 per ton. Owing to the absence of a guide familiar with the mine and lack of knowledge as to what portions of the material exposed should be classed as ore, no effort was made to systematically sample the deposit.

While attempts that have been made in the past to develop permanent mining operations at this locality have not been successful, yet, as there are probably large amounts of low grade ore available, the deposit warrants further investigation.

HART COUNTY

The gold deposits of this county occur in a region near Bowersville, a station on the Toccoa and Elberton branch of the Southern Railway. Mining operations have been confined in the main to small placer deposits along stream courses, though at one locality, the Brown mine, systematic underground mining was conducted a number of years ago on a vein deposit.

BROWN MINE.—This mine is situated about six miles south of Bowersville and about three miles from Royston. The principal mining operations that have been conducted here were carried on before the Civil war. The old works were in no condition for an examination and little or no surface exposures were noticed. According to reports, considerable drifting was done from the main shaft on an auriferous quartz vein and a stamp mill was erected at the time the mine was operated. Definite information as to the amount of gold secured was not available.

PLACER DEPOSITS.—Small placer deposits occurring along a branch have been mined on the Hays property on the Red Hollow public road about four miles from Bowersville and also on the adjoining Winter property. A little placer mining has also been done along the course of a small stream on the Bowers property about two miles from Bowersville on the public road to Royston.

FORSYTH COUNTY

FAVER MINE.—This mine is located in the western part of the county about twelve miles northwest of Duluth, a station on the main division of the Southern Railway. Placer mining has been conducted here along a small stream and some washing of surface material has also been done on a slope adjacent to the placer deposit. In addition to these operations vein prospecting has recently been in progress. Panning tests made at the time of visit from some fringes of unworked gravels near the branch yielded excellent results, the gold obtained being rather coarse. It is not probable, however, that sufficient unworked placer deposits occur at this locality to repay very extensive mining operations. More elaborate tests than could be undertaken when the property was examined would be necessary in order to determine whether any extended areas on the slopes adjacent to the placer would repay hydraulic mining. Some vein prospecting has been done in the vicinity of the placer, but at the time of visit, the operations had not been carried far enough to enable satisfactory conclusions to be drawn concerning the probable value of this class of deposits. A sample was taken for assay from a quartz vein having a rather flat dip that was exposed in a small cut or pit immediately adjacent to the placer. This sample on assay yielded no gold, but Mr. Faver reports that since the locality was visited additional prospecting has proved the presence of ore of a good grade.

SAWNEE MOUNTAIN PROPERTY.—Considerable mining for gold has been prosecuted at different times on Sawnee Mountain, a lofty ridge about four miles north of Cumming, the county seat. Several branches and gulches on the southeastern slope of the mountain have been extensively mined for placer gold, and, according to reports, a large amount of gold has been obtained. Numerous pits and shafts have been sunk in prospecting for valuable vein deposits, but up to the time of visit no veins seemed to have been located that warranted permanent mining operations. From some quartzose saprolite material in one of the prospect pits a little gold was panned when the property was examined and small amounts of gold were also secured from panning surface material at several points.

As no mining operations were in progress at the time of visit, it was difficult to draw conclusions as to the probable field at this locality for profitable surface mining in the future. In Bulletin No. 4-A, of the Geological Survey of Georgia, published in 1896, Mr. Francis P. King, who visited the property, says in speaking of some hydraulic mining that had been in progress at the locality under the direction of Messrs. Hampton and Herman, of Atlanta: "The results of their labor, however, tend to show, as conclusively as is possible on mineral properties of this character, that Sawnee Mountain can be worked with profit."

Mining and prospecting have been prosecuted at this locality over a considerable area. Some of the more important operations have been conducted on land lots 891 to 894 inclusive, and 909 to 911 inclusive, all in the 3rd district.

STRICKLAND MINE.—On lots 867 and 868, 3rd district, mining operations were conducted a number of years ago on auriferous quartz veins. The locality is close to the Cherokee county line and ore was hauled during the early mining operations to the Creighton, or Franklin, mine where it was milled. It is reported that considerable amounts of gold have been secured from this property.

PARKS AND FOWLER PROPERTY.—This property, in the western part of the county, is about nine miles from Cumming. Considerable prospecting was done here soon after the Civil war. Some years later additional mining operations were prosecuted, especially on lots 934 and 935, 3rd district. Nothing definite was learned as to the success that attended the work done at this locality.

OTHER PROPERTIES IN FORSYTH COUNTY.—Some prospect work has been done on the Lyons lot, 1259, 3rd district. Placer deposits have also been mined that occur along small streams tributary to Young Deer Creek in the southwest portion of the 14th district. About five years ago a rich pocket is reported to have been mined on lot 70, 2nd district, within a mile or so of Cumming. It is stated that $1,500.00 worth of gold was obtained.

CHEROKEE COUNTY

In addition to the auriferous deposits of the Dahlonega belt, which traverses Cherokee county, gold has been found at a few localities to the southwest of thisbelt. Several placer deposits have been mined which occur in the southwestern part of the county in the region of Little River. Some vein prospecting has also been done on lots 804 and 805, 2d district, in the southeastern corner of the county.

HABERSHAM COUNTY

Gold in isolated localities has been found at a few points in this county between the Dahlonega and the Hall county belts. The La Prade mine, on lot 135, 11th district, is located at the junction of two small streams tributary to Soque River. A placer deposit has been mined here that is reported to have yielded rich returns for the work that was done. It is stated that between twenty and thirty thousand pennyweights of gold were secured from an area of a few acres. Some large nuggets are also said to have been found at this locality. A limited amount of vein prospecting has been conducted in the vicinity of the placer.

A little prospect work has been done at several other localities between the two gold belts mentioned above, but no auriferous veins seem to have been located that warranted permanent mining operations.

GILMER COUNTY

Auriferous deposits occur at several localities in the eastern part of this county. The principal mining operations have been confined to placer deposits, although some limited mining operations have been prosecuted on auriferous veins.

WHITE PATH MINE.—This mine is situated near the Atlanta and Knoxville Division of the Louisville & Nashville Railway, and is about six miles northeast of Ellijay, the county seat. A placer deposit occurring here along a small creek has been worked at intervals for many years and according to the statements of reliable parties large amounts of gold have been secured. The greater portion of the area that has been mined occurs on the south side of the public road from Ellijay to White Path, on lots 270 and 288, 7th district, though work has also been done in dry hollows on the north side of the road. Some of the most productive areas were dry hollow or gulches on the southern edge of the placer. Two of these gulches, known as Austin's Bluff and Sprigg's Hollow, are especially noted for the amounts of gold obtained. According to reports, some of the largest nuggets that have been found in the State were discovered in working this placer. Two of these, found nearly simultaneously, are stated by reliable parties to have had an aggregate weight of about sixteen hundred pennyweights. Much of this placer has been worked over a number of times, mining operations having been inaugurated more than a half a century ago. Some nuggets are reported as having been found on a ridge to the south of the placer at an altitude of sixty or seventy feet above the stream bed and while the main portion of the placer has doubtless been exhausted, surface deposits of economic importance in hollows and on slopes in the vicinity of this mine may yet be discovered.

At time of visit some prospecting for veins was in progress on the slope of the ridge to the south of the placer. Some cross cuts had been made and a shaft sunk at a point near

Austin's Bluff. A temporary suspension of work prevented the shaft from being entered. Prospecting for auriferous veins in the region contiguous to the White Path mine seems never to have been carried on as systematically and thoroughly as the richness of the placer warranted. The gold in the placer may have been derived from numerous small stringers of auriferous quartz or from a zone of country rock that was partly silicified and impregnated with auriferous material. Such a condition would account for the apparent absence of distinct quartz veins in the region. Another possible explanation of the source of the gold may be found in the hypothesis that quartz veins of the ordinary type, occurring in that portion of the country rock that has been removed by denudation and erosion, supplied the gold to the placer, and pinched out at the present level of the land surface. Paleozoic rocks of sedimentary origin are found in the immediate vicinity of the mine.

A number of parties have a proprietary interest in the White Path mine. Messrs. J. A. Allen, of Ellijay, Ga.; W. A. Charters, of Gainesville, Ga.; and Judge G. F. Gober, of Marietta, Ga., all have an interest in lot 270, and Messrs. J. A. Allen, of Ellijay, Ga.; Samuel Tate, of Tate, Ga.; and the Enoch Faw Estate, of Marietta, Ga., have an interest in lot 188.

OTHER DEPOSITS IN THE VICINITY OF THE WHITE PATH MINE. Placer deposits have been mined to a limited extent along Little Turnip Town Creek, on the Holt property, a short distance south of the White Path mine. Below the junction of Big and Little Turnip Town creeks a broad expanse of valley land occurs and some years ago test pits were sunk and a ditch of considerable length constructed with the view of installing a hydraulic plant and mining these bottoms. For some reason the plans were never carried out. On lot 236,

10th district, about two miles west of the White Path mine, a limited amount of placer mining has been done along a small stream on the Whitaker property. The Turkey Pen mine, on lot 145, 7th district, is a few miles from the White Path mine. Washing of surface material on a hillside was done here a number of years ago. It is reported that at some points considerable amounts of gold was obtained. Some vein prospecting has also been carried on at the locality.

CARTECAY MINE.—The Cartecay mine, on lot 139, 6th district, is situated close to the Cartecay River and is about six miles east of Ellijay. Both vein and placer mining have been conducted at this locality. Old placer works are to be found for a considerable distance along the lower course of a small stream tributary to Cartecay River. The greater portion of these old works is now overgrown with trees and shrubs. The early returns are said to have been quite profitable and it is probable that the valuable portions of the placer were pretty well exhausted when the mining was done.

A limited amount of vein mining was conducted a number of years ago in the vicinity of the placer. A rich pocket was discovered that attracted considerable attention. A tunnel was afterwards driven along the strike of the vein for a short distance and a small stamp mill was erected and some milling done. No work has been in progress for a number of years and·nothing definite can be stated concerning the character of the vein.

A little placer mining has been done at several additional localities in the region of the Cartecay mine. Among others, the Reese, the Johnson and the Smith properties may be mentioned.

HALL COUNTY

The auriferous deposits to be described under the heading of isolated areas, occur in this county in the northern part and a few miles northwest of the Dahlonega gold belt. About the region of Murrayville, which is eleven miles north of Gainesville, several deposits are so related in position that they form a small belt with the characteristic northeast-southwest trend that might be designated as the Murrayville belt. Between this and the main Dahlonega belt, mining has been prosecuted at a few localities.

BIG JOE MINE.—This mine, named after Mr. Joseph Clements, the well known miner of Dahlonega, is situated on lot 61, 11th district, and is nine miles from Gainesville on the road to Murrayville. Systematic underground mining was conducted on an auriferous quartz vein at this locality about ten years ago. It is stated that the vein was worked to a depth of approximately sixty feet and drifting and tunneling was done. No surface exposures were noticed at time of visit and nothing can be stated from direct knowledge concerning the deposit. According to Mr. J. B. Clements, who worked the mine, the vein, where exploited, varied in thickness from eight inches to four feet and had an average thickness of from two and a half to three feet. A five-stamp mill was erected, when the mine was operated, which was later enlarged to ten stamps. Mr. Clements states that as the result of about five months' work something over five thousand dollars worth of gold was secured.

POTOSI MINE.—The Potosi mine is situated about eleven miles northwest of Gainesville and is on lot 85, 11th district. An auriferous quartz vein occurs at this locality that has been mined to a limited extent at intervals and by different parties through a long period of time. According to report, the deposit was worked for a couple of years by two English miners seventy-five years ago. The last mining operations of any importance were conducted about six years ago by Mr. Charles Kingsbury of Atlanta. It is stated that some very rich ore has been secured from this mine. As an examination of the property proved very unsatisfactory, owing to the fact that no mining was in progress, the following concerning the vein is quoted from Bulletin No. 4-A, of the Geological Survey of Georgia published in 1896:

"The vein, in which they are attempting to rediscover the rich chute, is about eighteen inches in thickness with a vertical dip and striking N. 50° W., cutting the country schists at a wide angle. It consists of a milky white quartz with a structure of parallelly arranged, interlocking acicular crystals. In the joints of the vein, made by the meeting of these crystals in their horizontal extension, free gold is sometimes, but rarely, found. In the ore body there is, however, no trace of sulphides."

ODOM MINE.—The Odom mine, on lot 111, 11th district, is about two miles west of Murrayville. Mining on a limited scale has been coducted here at intervals for many years. It is stated that some rich ore has been obtained at this locality. No work had been in progress for some years previous to the time of visit and an examination of the property, therefore, proved rather unsatisfactory. From Bulletin No. 4-A, of the Geological Survey of Georgia, published in 1896, the following description of the deposit, by Mr. Francis P. King, is quoted:

"The vein, upon which all have worked, consists of a system of quartz stringers intercalated in mica schist forming an ore body about three feet thick. This ore body lies on the southeast side of a hill near the east side of the lot. The sample taken by me for assay gave, in the laboratory of the Survey, a value in gold of $6.00 per ton."

SMITH PROPERTY.—Some prospect work was done on this property, which lies about a half of a mile west of Murrayville, several years ago. A number of shallow pits and cross cuts were made on several quartz veins.. It was not ascertained what success attended the work.

PARKS MINE.—The Parks mine, on lot 31, 11th district, is about four miles northeast of Murrayville. An auriferous quartz vein occurs here on which mining operations have been prosecuted at different periods. A limited amount of work was carried on at the locality about twenty years ago, at which time a small stamp mill was erected on the property. Later, in 1901, additional mining operations were prosecuted and a ten-stamp mill installed.

An open cut of considerable length that was made years ago is to be seen along the strike of the vein which cuts the country rocks at a wide angle. A shaft was sunk during the course of some of the operations from the bottom of this old cut near its centre. In 1901, a tunnel was driven from near the base of a hill for a distance of 300 feet striking the vein at about eighty feet from the surface. No mining has been done on this vein for a number of years and with the works in their present condition no attempt was made to secure samples for assay. Neither was anything very definite learned concerning the value of the ore that was taken out. It is reported that the vein is of large size and well defined.

OTHER ISOLATED LOCALITIES IN HALL COUNTY.—Some prospect work has been done on a quartz vein occurring on the A. M. Boggs property, lot 35, 11th district. Some vein prospecting has also been done near Cool Spring church, on lot 31, 10th district.

MURRAY COUNTY

A small isolated auriferous area occurs on the summit of Cohutta Mountain in Murray county. The locality is about four miles east of Chatsworth, a station on the Atlanta & Cincinnati Division of the Louisville & Nashville Railway. An interesting geological feature of the gold veins at this locality is their occurrence in the eastern edge of the Palae-ozoic area of Northwest Georgia, auriferous quartz veins being found in direct association with metamorphosed rocks of sedimentary origin.

COHUTTA MINE.—This mine, on lot 257, 26th district, is near the crest of Cohutta Mountain and about four miles east of Chatsworth. The existence of a small placer deposit along a branch a mile west of the Cohutta mine has been known for many years and a limited amount of placer mining was conducted on this deposit previous to the discovery of aurifer-ous veins. About eight years ago several gold bearing veins were located at the Cohutta mine and mining operations were inaugurated by the Cohutta Gold Mining Company, Mr. Paul Trammel of Dalton, Ga., being president. Several shafts were sunk and some open cuts made, and later a stamp mill was erected, but for some reason, after a little development work was done, operations were suspended and the mine has been idle for several years. During a visit to the property made soon after mining had commenced an opportunity was afforded to examine several auriferous quartz veins on which work was being prosecuted. One, known as the Little Mil-dred, had been exposed for a distance of about ninety feet in a shallow open cut. As seen there it showed a thickness of from four to ten inches and free gold was noticeable in some of the ore examined. The wall rock at this cut is a metamorphosed arkosic sandstone (description of rocks,

Chapter IV) apparently grading a short distance to the north-west into a conglomerate. A fine grained bluish schist or slate occurs immediately southwest of the cut. About two hundred yards east of the Little Mildred vein some work has been conducted on one or two other veins, but the exposures were not sufficient at that time to throw much light on their prob-able value. Some small specimens of ore were examined at this property that showed considerable amounts of free gold, but no very satisfactory conclusions were reached as to the probable extent of the deposits.

OTHER AURIFEROUS DEPOSITS IN THE VICINITY OF THE COHUTTA MINE.—As previously mentioned, a limited amount of placer mining has been done on a branch about a mile west of the Cohutta mine. Placer deposits are also said to occur along Chicken Creek, on lots 256 and 285. Some prospect work for veins has also been done on lot 229, known as the Stafford lot. At one point on this lot a quartz vein has been exposed for a depth of several feet in a prospect pit or shaft. A sample was taken from the vein in this pit, but on being assayed in the laboratory of the Survey it yielded no gold.

FANNIN COUNTY

Gold has been found in this county in commercial quantities in a limited area about twelve miles southeast of Blue Ridge, the county seat. The gold bearing region is traversed by Noontootly Creek along which some placer deposits occur. In addition to the deposits of the Noontootly Creek region a little placer gold has been found along a tributary of Sugar Creek at a point about a mile west of Blue Ridge.

RANTZE HILL MINE.—This mine on lot 285, 7th district, is situated close to Noontootly Creek. A placer deposit occur-ring here along a branch was mined before the Civil war, the

deposit forming a continuation of a placer deposit along
Noontootly Creek. On a hillside near the placer considerable
mining has been done at different times on auriferous quartz
veins. Several old tunnels, shafts and open cuts are to be
seen at the locality on the side of the hill. A number of small
quartz veins composed of granular quartz containing apparently only small amounts of sulphides occur here together
with some larger veins which show a thickness in places of several feet. The mining operations seem to have been confined
principally to the smaller veins. In 1901, the mine was worked
for about a year by the Rantze Hill Gold Mining Company,
who installed a Huntington mill at the locality. It could not
be ascertained what gold was secured. Mr. Willis Garrett of
Newport, Ga., exhibited some specimens of ore that had been
taken from the mine that were quite rich in free gold.

OTHER AURIFEROUS DEPOSITS NEAR THE RANTZE HILL MINE.—
Placer deposits on Noontootly Creek and tributaries near the
Rantze Hill mine have been worked at different times and
it is reported that considerable amounts of gold have been
secured. Some vein prospecting has also been done on lot
321, 7th district, and on lot 294 in the same district. The Mc-
Cloud mine is a placer deposit occurring along a small
branch on lot 106, 6th district. The deposit has been mined
for a distance of about a hundred and twenty-five yards along
the course of the branch and it is reported that over three
thousand dollars worth of gold was secured.

INDEX

www.ingramcontent.com/pod-product-compliance
Lightning Source LLC
Chambersburg PA
CBHW031806190326
41518CB00006B/216